Oxide Dispersion Strengthened Refractory Alloys

Oxide Dispersion Strengthened Refractory Alloys

Anshuman Patra

CRC Press
Taylor & Francis Group
Boca Raton London New York

CRC Press is an imprint of the
Taylor & Francis Group, an **informa** business

First edition published 2022
by CRC Press
6000 Broken Sound Parkway NW, Suite 300, Boca Raton, FL 33487-2742

and by CRC Press
2 Park Square, Milton Park, Abingdon, Oxon, OX14 4RN

© 2022 Anshuman Patra

CRC Press is an imprint of Taylor & Francis Group, LLC

Reasonable efforts have been made to publish reliable data and information, but the author and publisher cannot assume responsibility for the validity of all materials or the consequences of their use. The authors and publishers have attempted to trace the copyright holders of all material reproduced in this publication and apologize to copyright holders if permission to publish in this form has not been obtained. If any copyright material has not been acknowledged please write and let us know so we may rectify in any future reprint.

Except as permitted under U.S. Copyright Law, no part of this book may be reprinted, reproduced, transmitted, or utilized in any form by any electronic, mechanical, or other means, now known or hereafter invented, including photocopying, microfilming, and recording, or in any information storage or retrieval system, without written permission from the publishers.

For permission to photocopy or use material electronically from this work, access www.copyright.com or contact the Copyright Clearance Center, Inc. (CCC), 222 Rosewood Drive, Danvers, MA 01923, 978-750-8400. For works that are not available on CCC please contact mpkbookspermissions@tandf.co.uk

Trademark notice: Product or corporate names may be trademarks or registered trademarks and are used only for identification and explanation without intent to infringe.

ISBN: 978-1-032-06165-8 (hbk)
ISBN: 978-1-032-06166-5 (pbk)
ISBN: 978-1-003-20100-7 (ebk)

DOI: 10.1201/9781003201007

Typeset in Times
by Newgen Publishing UK

Contents

Preface ix
Author xi
Acknowledgments xiii

Chapter 1   Refractory Metals: Fundamentals and Applications 1

    1.1   Definitions 1
    1.2   Structure 1
    1.3   Properties at a Glance 2
           1.3.1   Properties of Tungsten 3
           1.3.2   Properties of Molybdenum 3
           1.3.3   Properties of Niobium 4
           1.3.4   Properties of Rhenium 4
           1.3.5   Properties of Tantalum 5
           1.3.6   Properties of Iridium 6
           1.3.7   Properties of Ruthenium 6
           1.3.8   Properties of Osmium 7
           1.3.9   Properties of Hafnium 8
    1.4   Applications 9
           1.4.1   W and W Alloys 9
           1.4.2   Mo and Mo Alloys 11
           1.4.3   Nb and Nb Alloys 12
           1.4.4   Re and Re Alloys 13
           1.4.5   Ta and Ta Alloys 13
           1.4.6   Ir and Ir Alloys 14
           1.4.7   Ru and Ru Alloys 14
           1.4.8   Os and Os Alloys 14
           1.4.9   Hf and Hf Alloys 15
    1.5   Major Limitations with Respect to Applications 15
    1.6   Basics of Nano-Structured Materials: Experimental and Thermodynamic Modeling-Based Evidence 19
    1.7   Inferences 25
    References 26

Chapter 2   Principles of Oxide Dispersion Strengthening 35

    2.1   Characteristics of Oxides 35
    2.2   Criteria for the Selection of Oxides 39
    2.3   Dispersion Strengthening 39
    2.4   Mechanism of Oxide Dispersion Strengthening 41
    2.5   Suitability of Nano-Oxides Over Micron-Sized Oxide Dispersion 43

	2.6	Effectiveness of Dispersion Strengthening by Oxides............44
	2.7	Inferences..47
	References..47

Chapter 3 Processing Fundamentals 51

    3.1 Conventional Melting-Casting..51
    3.2 Limitations of the Melting-Casting Route 53
    3.3 Mechanical Alloying..53
    3.4 Rapid Solidification 56
    3.5 Severe Plastic Deformation..58
        3.5.1 Equal Channel Angular Pressing................................59
        3.5.2 Multidirectional Forging 60
        3.5.3 High Pressure Torsion 61
        3.5.4 Accumulative Roll Bonding..62
        3.5.5 Twist Extrusion...64
    3.6 The Sol-Gel Method 65
    3.7 Chemical Vapor Deposition 65
    3.8 Physical Vapor Deposition 67
    3.9 Laser Ablation..68
    3.10 Inferences..68
    References..69

Chapter 4 Synthesis of Oxide Dispersion Strengthened Refractory Alloys.........77

    4.1 Effects of Oxide Dispersion on the Processing of Refractory Alloys..77
    4.2 Phase Evolution in Oxide Dispersion Strengthened Alloys...78
    4.3 Microstructural Characterization 81
    4.4 Effect of Process Parameters on Microstructural Development...83
    4.5 Inferences..86
    References..86

Chapter 5 Consolidation Methods 91

    5.1 Conventional Pressureless Sintering..91
    5.2 Spark Plasma Sintering...94
    5.3 Microwave Sintering...98
    5.4 Hot Isostatic Pressing..100
    5.5 Hydrostatic Extrusion 101
    5.6 Powder Forging..103
    5.7 Sinter Forging 105
    5.8 Shock-Based Consolidation 106
    5.9 High Pressure Sintering..108
    5.10 Inferences..110
    References..110

Contents

Chapter 6 Densification of Consolidated Products 117
    6.1 Mechanism of Densification 117
    6.2 Strategy to Enhance Densification 123
    6.3 Role of Densification in Final Applications 125
    6.4 Inferences 127
    References 127

Chapter 7 Mechanical and Wear Behavior 131
    7.1 Hardness 131
    7.2 Strength and Toughness 132
    7.3 Mechanism of Deformation at Ambient and Elevated Temperatures 134
    7.4 An Approach to Improve Fracture Toughness 138
    7.5 Wear Properties 142
    7.6 Inferences 144
    References 145

Chapter 8 High Temperature Oxidation Characteristics 149
    8.1 Oxidation of Refractory Alloys 149
    8.2 Effect of Oxide Dispersion on the Oxidation of Refractory Alloys 151
    8.3 Factors Influencing the Oxidation of Oxide Dispersion Strengthened Alloys 155
    8.4 Methods to Enhance Oxidation Resistance 157
    8.5 Inferences 160
    References 161

Chapter 9 Application-Oriented Strategy 165
    9.1 Properties that Influence the Applications 165
    9.2 Improvement of Applicability 168
    References 169

Chapter 10 Future Potential 171
    10.1 Trend Toward Improvement of Sustainability 171
    10.2 Focus to Extend the Application Areas 173
    References 177

Index 181

Preface

The focus of this book is the development of oxide dispersion strengthened refractory alloys, their physical and mechanical properties and high temperature oxidation behavior. Several innovative fabrication methods for complex, near net shape fabrication assist in sustainable product fabrication. Dispersion of oxides stimulates the high temperature application of refractory alloys. Designing alloys by adding nano-oxides, controlling the grain size distribution, regulating the consolidation parameters (time, temperature, heating rate, atmosphere, pressure), the volume fraction of dispersed oxides, the extent of the dispersion homogeneity, and the dispersion sites are some of the key aspects to improve the final properties. Extensive research on the fabrication of refractory ODS alloys has been carried out to date, however, more stringent pollution norms, property requirements, and bulk-scale production, all require state-of-the-art manufacturing technology to be embraced.

The book is divided into 10 chapters detailing processing methods, oxide dispersion, properties and sustainable applications.

Chapter 1 illustrates the structure, properties and applications of refractory metal alloys along with their application-based challenges. The fundamentals of nanostructured materials are also discussed in this chapter. The features of oxides and guidance in selecting oxides to add in refraction metals are presented in Chapter 2. The chapter also includes the fundamental understanding of the mechanisms of dispersion strengthening, various aspects of nano-oxide dispersion and the effectiveness of oxide dispersion. Chapter 3 presents various processing methods of refractory metals and alloys. The importance of the methods and their challenges are also described. The impact of oxide dispersion during the processing of refractory alloys, the phases, the microstructural investigation, the role of various process parameters on the microstructure of the processed alloys are discussed in Chapter 4. Chapter 5 includes different consolidation techniques and the influence of various parameters on the consolidated microstructure. The principle of densification during consolidation, methods to enrich density, and the impact of density on the application of refractory ODS alloys are described in Chapter 6. Chapter 7 explains the mechanical properties (hardness, strength and toughness), the principle of ambient and elevated temperature deformation, fracture toughness enhancement and wear behavior of refractory ODS alloys. The oxidation characteristics of refractory alloys and the importance of oxide dispersion and the process parameters with respect to oxidation, and enrichment of oxidation resistance are explained in Chapter 8. Chapter 9 presents the properties related to the application of oxide dispersion strengthened refractory alloys and Chapter 10 describes sustainable alloy development with an emphasis on widening the application spectrum.

The book provides a comprehensive introduction to refractory alloys, showing the influence of oxide dispersion strengthening in refractory alloys, highlighting the different fabrication parameters, regulating the microstructure to achieve the desired sustainability in applications. The book will be useful for researchers investigating the field of high temperature refractory alloys and oxide dispersion strengthening.

Author

Anshuman Patra is an assistant professor in the Department of Metallurgical and Materials Engineering, National Institute of Technology Rourkela, Odisha, India. He has 5 years industrial experience in the ferrous and aluminum industries. He has won several awards and accolades, including a university medal for 1st position in the Department of Metallurgical and Materials Engineering during his undergraduate study, the Indranil Award from the Mining, Geological & Metallurgical Institute of India, and Zero Abnormality monitoring and Kaizen awards during his tenure in the aluminum industry. He has published more than 20 research publications in international journals, has been the author and co-author of two books and has also delivered several invited talks. He is on the review board of several reputed journals and has also been certified as an outstanding reviewer by several Elsevier journals. He is also a life member of the Indian Science Congress Association, the Powder Metallurgy Association of India, and the Indian Institute of Metals. His research interests include high temperature metals and alloys, sintering, powder metallurgical processing, oxide dispersion strengthened refractory alloys.

Acknowledgments

I would like to acknowledge the inspiration provided by my son Aayan, wife Bijeta, sister Sudeshna and my parents for their motivation to write this book. The support of the National Institute of Technology Rourkela, Odisha, India, is also acknowledged.

1 Refractory Metals
Fundamentals and Applications

1.1 DEFINITIONS

Refractory metals are defined as metals with a high melting point and, in general, their lower limit of melting point is 2000°C. Refractory metals have varied applications, such as in aerospace, nuclear fields, defense systems, chemicals, and electronics [1]. In the following sections, the structure, the properties, the applications, and the limitations of the refractory metals such as Tungsten (W), Molybdenum (Mo), Niobium (Nb), Rhenium (Re), Tantalum (Ta), Iridium (Ir), Ruthenium (Ru), Osmium (Os), and Hafnium (Hf) will be discussed.

1.2 STRUCTURE

The crystal structure and its interatomic interaction significantly influence the properties of refractory metals. Among the refractory metals, W, Mo, Nb, Ta are body-centered cubic (BCC), Re, Ru, Os, Hf are hexagonal closed packed (HCP), Ir is a face-centered cubic (FCC). W, Mo, Nb, Ta are categorized as key refractory metals and Re, Ru, Os, Ir, are trivial refractory metals. The atomic packing factor of BCC is 0.68 whereas for HCP and FCC, it is 0.74. Though W is classified as BCC metal, however, it possesses the highest density. The bond strength of the refractory metals is related to the melting temperature, the ionization potential, self-diffusion, the Debye temperature, etc. Refractory metals belong to the transition metals. HCP transition metals show stability in the range $1 \leq n_d \leq 2$ and $6 \leq n_d \leq 7$, for FCC, it is $8.5 \leq n_d \leq 10$ (n_d: d-occupation). All transition refractory metals with a BCC structure are stable for every further n_d [2, 3]. W possesses an electronic configuration of [Xe] $6s^2 4f^{14} 5d^4$ [4]. The Fermi energy in W is between the bonding and antibonding positions. The unsaturated resilient covalent bonds among the $5d$ orbitals lead to stabilization of the BCC structure [5]. The d shell configuration in Mo, Nb, Re, Rh, Ru, Os, Hf is $4d^5$, $4d^4$, $5d^5$, $4d^7$, $5d^5$, $5d^2$ respectively. The half-filled/partly filled d orbitals result in a strong bond energy which results in high melting points of refractory metals. The d shell configuration also influences the interatomic bonding and the cohesive energy. The half-filled d orbitals are related to most interatomic bonding and result in the highest density and cohesive energy of the refractory metals [6].

The cohesive energy or the bonding energy of the metals is the basis for an understanding of the bulk modulus and elastic properties of metals. The cohesive energy of W is 8.90 ev/atom, and is maximum compared to Mo (6.82 ev/atom), Nb (7.57 ev/atom), Ta (8.10 ev/atom) [7]. The BCC structure of W can transform to FCC

TABLE 1.1
Lattice Constant of Refractory Metals

Refractory metals	Lattice constant (Å)	
W	3.16379	
Mo	3.1470	
Nb	3.300	
Re	2.7615	4.4566
Ta	3.30256	
Ir	3.8390	
Ru	2.7058	4.2816
Os	2.7338	4.3195
Hf	3.198	5.061

Sources: [12–14].

or HCP by electronic stimulation within the electronic temperature range of 1.7–4.3 ev [8]. However, it is important to understand the stability of the structures. Considering the total energy of W with respect to all atomic volumes, BCC W exhibits the lowest energy compared to FCC-W and HCP-W [9]. BCC W possesses the lowest volume/atom and its bonding is highest among other structures. Therefore, BCC-W shows the maximum stability. Similarly, Mo is BCC at normal room temperature and pressure. The phase transformation of Mo from BCC to FCC occurs at elevated pressure (~700 GPa), as presented by the density function theory-based simulation study [10, 11].

The lattice constant of various refractory metals (W, Mo, Nb, Re, Ta, Ir, Ru, Os, Hf) is presented in Table 1.1.

The interatomic strength is influenced by the interatomic distance and the electrostatic force among the atoms (F) is related to the interatomic distance (r) as:

$$F \propto \frac{1}{r^2} \quad (1.1)$$

The interatomic bond is highest in the metals of the VIA group in the periodic table because of the outer electron contribution in the s-p-d loop [15].

1.3 PROPERTIES AT A GLANCE

The main reason for selecting refractory metals compared to other structural metals is the melting point. The high melting point of refractory metals facilitates high temperature applications. However, several other properties of refractory metals, including high strength, elevated density, low thermal expansion coefficient, and low electrical and thermal conductivity, are of prime importance. The low electrical conductivity is due to the conduction electron's s-d transition and synergy with lattice vibrations [15]. The above-mentioned properties are application-oriented. During exposure at high temperatures, the deformation and oxidation resistance of refractory metals are

Refractory Metals: Fundamentals 3

of immense importance in order to improve their sustainability. The properties of W, Mo, Nb, Re, Ta, Ir, Ru, Os and Hf are discussed in the following sections.

1.3.1 Properties of Tungsten

W possesses a melting point of 3410°C, a density of 19.3 gm/cm^3, ultimate tensile strength (UTS) at 1000°C: 350 MPa, vapor pressure at 2226.85°C is 0.009 MPa, coefficient of thermal expansion at room temperature: 4.6 x 10^{-6}/K, room temperature thermal conductivity of 155 W/mK along with low sputtering yield, and low tritium retention [16, 17]. The elevated temperature strength of W reduces with an increase in the deformation temperature at fixed strain rate [18]. W also exhibits a low rate of sputtering, minimal enthalpy preservation, and minimum swelling [19]. The room temperature fracture toughness value of pure W is ~ 5–10 MPa$\sqrt{m}$ and the value increases to 20–25 MPa$\sqrt{m}$ at temperature >300–350°C [20–22]. Polycrystalline W displays noticeable elongation of 53% at 400°C against room temperature elongation of 0–0.3% [23]. Creep behavior has immense significance during high temperature structural applications of W. Reports suggest that creep deformation of W can start from 0.3T_m [24, 25]. The recrystallization and creep behavior of W are presented in studies [26–31] and it is reported that the activation energy decreases with a decrease in temperature [27]. The stability of a component at an elevated temperature is also influenced by its oxidation behavior. The oxidation rate of W at 1200°C is 200 mg/cm^2/h [16]. The oxide scale formed in W is adhesive at oxidation temperature <650°C, volatilization of the oxide scale occurs between 775–1250°C and the oxide layer is absent above 1250°C [32].

1.3.2 Properties of Molybdenum

Mo shows a melting point of 2620°C, a density of 10.2 gm/cm^3, UTS at 1000°C is 140–200 MPa, vapor pressure at 2226.85°C is 80 MPa, coefficient of thermal expansion at room temperature is 4.9 x 10^{-6}/K, and room temperature thermal conductivity is 142 W/mK [16]. The low thermal expansion coefficient of Mo supports dimensional stability and reduces the chances of cracking [33]. Appreciable thermal conductivity also facilitates reduced stress generation during heat treatment. The literature indicates that commercially pure Mo exhibits a UTS of 534 MPa, percentage elongation to failure of 13% and fracture toughness of 24.2 ± 2.3 MPa$\sqrt{m}$ at room temperature [34]. The creep strength of commercially pure Mo decreases with an increase in temperature with the temperature window of 1326.85–1826.85°C and 5% creep strain is achieved in Mo in < 3 days at 3.45 MPa load and 1676.85°C temperature [35]. Cho et al. have shown that the fracture time of pure Mo sheet enhances at reduced load and temperature, owing to the formation of secondary recrystallized grains [36].

The oxidation rate of Mo at 1200°C is quite elevated compared to other refractory metals (6000 mg/cm^2/h) [16]. Oxides of Mo volatilize to some extent at 600°C and 76 torr (0.01 MPa) O$_2$ pressure, followed by melting at 795°C [37]. Parabolic rate law is observed at 500°C and above 500°C, the linear rate law is followed with respect to oxidation of Mo [38]. Evaporation of MoO_3 appears at 650°C whereas destructive oxidation occurs at 725°C. Therefore, at elevated temperatures, there is catastrophic oxidation which significantly consumes the Mo and minimizes the durability of Mo-based components.

1.3.3 Properties of Niobium

The melting point of Nb is 2470°C, density is 8.6 g/cm³, UTS 90–120 MPa (at 1000°C), vapor pressure 5.3 MPa (at 2226.85°C), coefficient of thermal expansion at room temperature is 7.3 x 10^{-6}/K, and room temperature thermal conductivity for Nb is 53 W/mK [15]. Nb is ductile at room temperature and therefore deformable [39, 40]. The creep ductility of Nb is >50% with minor occurrence of void or intercrystalline cracking [41]. The activation energy (Q_c) for creep is expressed as [42]:

$$Q_z = KT^2 \left(\partial \ln \varepsilon_i / \partial T\right) / \sigma, Z \quad (1.2)$$

where ε_i is the creep rate, T is temperature (in K), σ is applied stress, Z is factor, representing constant microstructure.

For temperatures <$0.45T_m$ (T_m: melting temperature), $Qc < Q_{self\,diffusion}$. In this case, creep is regulated by cross-slip, pipe/grain boundary diffusion.

For temperatures > $0.5T_m$, $Qc = Q_{self\,diffusion}$, creep is diffusion controlled.

A report states that 100 h stress rupture study of Nb shows several percent primary creep, 10^{-6}/s of steady state creep rate and tertiary regime possess ≥1/3rd of overall test life [41].

Nb shows cleavage fracture on (001) and (110) at temperature < $0.07 T_m$. The fracture occurs by cracking along the cleavage/grain boundary [43]. With an increase in temperature, plastic deformation becomes evident and fracture occurs through ductile mode. Initiation of creep and transgranular fracture take place at temperature > $0.3 T_m$.

The oxidation of Nb beyond 500°C is quite intense. Various oxides such as NbO, NbO_2, Nb_2O_5 appear due to the oxidation of Nb at different temperature regimes. Nb_2O_5 forms as a result of heating Nb in air above 400°C which possesses high metal volume: oxide volume [44]. The volume ratio of $Nb_{0.98-1.00}$ oxide is 1.37 which increases to 1.87 for NbO_2 and 2.68 for $NbO_{2.43-2.50}$. The increase in the ratio leads to the development of stress and results in the bursting of the oxide scale [45]. The oxidation rate of Nb is 200 mg/cm²/h at 1200°C which is significantly less than Mo [16]. NbO_2 is the main volatile constituent compared to Nb_2O_5 with respect to 1 atm O_2 pressure [32]. Oxidation resistance of Nb is quite low above 1512°C as Nb_2O_5 appears as a liquid on the NbO_2 layer.

1.3.4 Properties of Rhenium

A high melting point of 3180°C, a density of 21 g/cm³ at 25°C, UTS at a high temperature (1000°C) of 800 MPa, vapor pressure of 0.17 MPa (at 2226.85°C), room temperature thermal expansion coefficient of 6.7 x 10^{-6}/K, room temperature thermal conductivity of 71 W/Mk, electrical resistivity of 19.3 μohm cm at 20°C are some of the properties of Re [46, 16]. The room temperature tensile strength is 1172 MPa and percentage elongation is 35% [47]. Rhenium can be workable at room temperature owing to sufficient ductility. The percentage elongation also depends on the fabrication method. Research has been conducted on the creep behavior of Re fabricated by powder metallurgy and electron beam melting. The percentage elongation to rupture for electron beam melted and powder metallurgically fabricated Re has been reported as 22% and 4% respectively at 1204.4°C and stress of 138 MN/sq.m [48]. Enhanced percentage elongation in electron beam melted Re results in extended rupture life. Rhenium possesses an appreciable recrystallization temperature of 1625°C and therefore exhibits enhanced creep resistance [47]. The life to rupture of Re is superior till 2800°C accompanied by elevated stress compared to W and exhibits less distortion with respect to extensive temperature fluctuations than W [47] and Mo. The oxidation rate of Re (2000 mg/cm^2/h at 1200°C) is 10 times higher than W and Nb but one-third higher than Mo [16]. Re shows reduced oxidation resistance due to the formation of low melting point and volatile oxides. Re_2O_7 oxide forms at 290°C but melts at 297°C and volatilizes > 635 K (~362°C) [49]. Oxides with a high sublimation rate consume the base metal and contribute to poor oxidation resistance. According to the literature, collapse of the surface is evident in Re prepared by chemical vapor deposition (CVD) and powder metallurgy along with cold rolling and recrystallization annealing [49]. The literature also indicates that a coarser grain size and the intense texture of the basal plane minimize the oxidation of Re.

1.3.5 Properties of Tantalum

The rare earth metal Ta shows a melting point of 3000°C, 16.6 g/cm^3 density at 20°C, elastic modulus at 20°C is 186 GPa, specific heat at 20°C of 139.1 J kg^{-1} K^{-1}, thermal expansion coefficient of 6.5 x 10^{-6}/K, lower room temperature thermal conductivity (54 W/mK) compared to W, Mo, Re, Ir, but marginally higher than Nb, UTS (at 1000°C) of 140 MPa and vapor pressure of 0.11 MPa at 2500 K (2226.85°C) [16, 50]. The hardness of pure Ta is 0.4 GPa and it increases with an increase in oxygen concentration [51]. Additionally, reduced work hardening rate and elongation of 40% are observed in Ta at room temperature [52]. Grain boundary sliding is evident during the creep deformation of Ta. The creep fracture occurs by transcrystalline mode with high fracture elongation [41]. The effect of increased Ta addition from 0.06 wt.% to 0.14% wt.% on the creep rupture property of ferritic-martensitic steel with a constant presence of W is a reduction in creep rupture strength and ductility [53]. The report also illustrates that increased addition of Ta hastens the start of tertiary creep.

Voitovich et al. have studied the elevated temperature oxidation behavior of Ta with varied purity [54]. During oxidation between 500–800°C of Ta, the formed oxide is Ta_2O which transforms to β-Ta_2O_5 till 1360°C [54, 55]. The formation tendency of β-Ta_2O_5 enhances with higher impurity. The parabolic oxidation law is prevalent between 900–1000°C and moves to the linear rate at an elevated temperature [55]. The linear rate with the formation of porous Ta_2O_5 is associated with an increased Pilling-Bedworth ratio of 2.5 and intense compressive stresses which enhance the cracking behavior of the oxide scale [55–57]. Report evidence shows the oxidation rate of Ta at

1200°C is identical to W and Nb [16]. The oxidation rate is also influenced by oxygen pressure which is also temperature-dependent. As the oxygen pressure increases, the weight gain also increases, however, at low temperatures, the high oxygen pressure has less effect on oxidation [56].

1.3.6 Properties of Iridium

The melting point of Ir is 2445°C, the density is higher (22.6 g/cm^3) compared to W, UTS of 330 MPa at 1000°C [16]. The thermal expansion coefficient is (6.8 x 10^{-6}/K) higher compared to W, Os, Mo, Hf, Ru and less than Nb at room temperature and the room temperature thermal conductivity is 147 W/mk [16]. It also has a reasonably high Young's modulus (500–650 GPa), and the microhardness is 2500–3000 MPa in the annealed condition and 6300–6500 MPa when 10–15% deformation is achieved [58]. The percentage elongation of the rupture of Ir is enhanced from 6.8% at 20°C to 18% at 800°C and the Vickers microhardness at 20°C is 200 HV(1.96 GPa) which reduces to 97 HV (0.95 GPa) at 1000°C [59].

The stress rupture strength of pure Ir decreases with the increase in time to rupture. Study of stress rupture strength between 1650–2300 °C shows that the stress rupture strength enhances with a decrease in temperature [60]. Ir fractures in a ductile manner during a high loading condition and fractures in an intercrystalline manner during extended rupture time and aa low loading condition. Rupture strain till 100% at elevated loads and maximum strain percentage at 2100°C have been reported in the literature [60]. Mehan et al. have also reported that the failure mode of Ir beyond $0.5T_m$ is partly intergranular and the minimum creep rate is associated with high rupture time and low stress condition [61].

IrO_2 oxide forms as a result of Ir oxidation above 400°C. Above 1100°C, dissociation of IrO_2 occurs [62]. Oxidation of Ir higher than 1100°C evidences the presence of volatile oxides IrO_3 and IrO_2 in the form of gas [63]. Other reports suggest that oxidation of Ir sheet at 1000°C exhibits non-significant weight change till 50 h [64]. According to Semmel, the oxidation rate of Ir at 1200°C is considerably smaller (2 mg/cm^2/h) compared to W, Mo, Nb, Re, Ta, Ru, Os [16]. Zhu et al. have investigated the oxidation behavior of an Ir-coated Re rod and demonstrated that, under high oxygen partial pressure (~2 × 10^4 Pa), oxidation of Ir directly results in collapse of the coating [65].

1.3.7 Properties of Ruthenium

Ru has a melting point of 2310°C, 12.4 g/cm^3 density, and room temperature thermal expansion coefficient of 5.1× 10^{-6} /K [16]. The UTS of Ru at 20°C is 500 MPa which decreases to 300 MPa at 750°C and the percentage elongation increases from 3% to 15% with an increase in temperature from 20°C to 750°C [59, 66]. The hardness varies between 250–500 HV at 20°C and 160–280 HV at 600°C with respect to various planes [66]. The thermal expansion coefficient of polycrystalline Ru changes from 8×10^{-6} to 8.8×10^{-6}/K with an increase in temperature from 623°C to 823°C [59]. The result is an indication of enhanced thermal stability at elevated temperatures.

The investigation of creep deformation behavior of pure Ru at elevated temperatures is limited. Song et al. have studied the increasing effect of Ru on creep behavior of single crystal Ni superalloys at 1140°C [67]. According to the study, the increasing Ru content from 2.5 wt.% to 3.5 wt.% enhances the creep rupture life of the alloy by hindering the generation of microvoids. Guoqi et al. have also indicated that the creep life time improved by 1.96 times by the addition of Ru in a single crystal Ni-Re alloy owing to the reduction of dislocation sliding and cross-sliding [68].

During oxidation in the air, initially Ru forms the RuO_2 scale of high density and with increasing temperature up to 1000°C; however, with increasing temperature at 720°C, RuO_4 is partly formed, and at elevated temperature RuO_3 appears [69]. According to the report, the oxidation rate of Ru at 1200°C is 10 mg/cm²/h, and the rate is much less compared to W, Mo, Nb, Re, Ta, and Os [16]. RuO_3 and RuO_4 are volatile oxides and enhance the oxidation. The partial vapor pressure of RuO_3 oxide increases from 500×10^{-5} atm at 1300°C to 4000×10^{-5} atm at 1500°C and aggravated the volatile oxide formation [69]. The partial pressure of RuO_4 at 600 K is almost 10^{-6} bar (0.986×10^{-7} atm) and the high volatility of RuO_4 is attributed to less binding energy due to impeccable saturation of atoms and structural symmetry [70, 71]. There may be a chance of finding volatile RuO_4 in the containment region of nuclear power plants and a probable accident in a highly oxidizing environment [71].

1.3.8 Properties of Osmium

The melting point is 3045°C, room temperature density is 22.6 g/cm³, room temperature thermal expansion coefficient and room temperature thermal conductivity (polycrystalline Os) of Os are 2.6×10^{-6} /K, 88 W m⁻¹ K⁻¹ respectively [16, 72]. Os shows hardness variation of 2.9 GPa–6.7 GPa at 20°C along with room temperature brittleness based on crystal orientation and the variation drops to 1.3 GPa–3.9 GPa at 1200°C [59]. The reported value of tensile strength in annealed Os is 1000 N/mm² (1000 MPa) [73]. The Young's modulus of Os at room temperature is 560 GPa which drops to around 500 GPa with an increase in temperature to 700°C and reported shear modulus is 220 GPa [51, 72]. Os (paramagnetic) also exhibits molar magnetic susceptibility of 138×10^{-6} cm³ mol⁻¹ at 22°C [72] and specific heat capacity of 130.9 J kg⁻¹ K⁻¹ at 298.15 K [51].

The oxidation study of Os at ≥ 1000°C is quite limited. OsO_2 on Os occurs at room temperature [69]. The oxides formed during oxidation of Os in the air are volatile OsO_4 or OsO_3, as reported by Grimley et al., between 825–1475°C with respect to oxygen pressure of $1.2 \times 10^{-5}/3.5 \times 10^{-5}$ atm [74]. Beyond 1425°C, OsO_2 oxide of extremely small intensity has been reported. The tentative linear weight loss rate of Os at the temperature and pressure of 1400°C and 1 atm respectively is 1.2×10^3 mg/cm²/h in comparison to 1.35×10^3 mg/cm²/h at 1000°C [75], whereas other reports indicate that the oxidation rate of Os at 1200°C is 800 mg/cm²/h, the third highest among other refractory metals (Mo> Re> Os>W= Ta=Nb>Ru>Ir>Hf, W, Ta; Nb possesses the same oxidation rate at 1200°C). The mass loss of Os within the time range of 15 min changes almost constantly at 600°C to highly linear at 1400°C [76]. The weight loss at elevated temperatures is attributed to high vapor pressure of toxic

OsO_4 [69]. The investigated oxide surface of Os at 600°C is extremely oxidized in some of the surfaces and at 1000°C shows several fractures [75].

1.3.9 Properties of Hafnium

Hf is denoted as a group 4 transition metal according to IUPAC 1988 and has a melting point of 2230 °C, density of 13.3 g/cm^3, thermal expansion coefficient at room temperature 5.9×10^{-6}/K, and room temperature thermal conductivity of 22 W/mK [16]. The allotropic transformation temperature of Hf is 1760°C and the crystal structure at room temperature is HCP (α) which transforms to BCC (β) above 1760°C [77]. The specific heat and electrical resistivity (at 25°C) is 146 J kg^{-1} °C^{-1}, 35 µΩ cm and Young's modulus, the shear modulus at 20°C is 140 GPa, 54.2 GPa [77]. Hafnium can be forged or hot rolled with a rolling temperature of 900–1000°C and in annealed form Hf exhibits a hardness of 130 HV corresponding to an oxygen content of 40 ppm [78–82]. The hardness and the UTS of HF depend on the oxygen content and the room temperature properties improve with an increase in the oxygen concentration. The UTS is 350 MPa,, 0.2% yield strength (YS) is 140 MPa, and the percentage elongation of Hf in an annealed condition with < 50 ppm oxygen is 38%. Hf possesses c/a ratio less than the ideal c/a ratio (1.633) of HCP metals and therefore slip deformation occurs in the prism planes [83]. However, in a contrary study on stacking fault energy, it is shown that the basal plane is the favored plane in Hf, owing to lower general stacking fault energy compared to the prism plane [84].

Very few reports on creep properties of pure Hf are available. Hf is effective in improving the creep strength and creep rupture life in superalloys [77]. The variation between stress versus minimum creep rate at 399°C (in a longitudinal direction) of forged (1095°C), rolled (980°C) and vacuum annealed (900°C, for 20 h) Hf shows a linear increasing trend [85]. In addition, a creep test in the longitudinal direction of forged (899°C), rolled (927°C) and heat-treated at 927°C (15 min, Argon) Hf shows higher stress at minimum creep rate between 10^{-6} to 10^{-5} in/in-h compared to an Hf sample tested at 399°C. Cross-slip nucleation/dislocations intersections are the probable method of creep in Hf at this temperature [85].

The oxygen solubility in α-Hf drops with a decrease in temperature (from 22 at.% at 2000°C to 20 at.% at 700°C) and moreover oxygen stabilizes the α-Hf and the conversion temperature to β-Hf increases [86]. The literature reports that the oxidation of Hf in the temperature regime of 350–1200°C follows a logarithmic, parabolic, linear profile and as the temperature increases, the logarithmic rate converts to the parabolic rate [87]. It is also indicated in that paper that oxidation proceeds through diffusion via lattice and grain boundary. Voitovich et al. have stated that oxidation of remelted Hf powder increases more intensely at 900°C compared to 800°C, 950°C, 1000°C, 1100°C and 1200°C [88]. The oxidation product HfO_2 (with a monoclinic structure) possesses a PBR (Pilling-Bedworth ratio) of 1.5 and leads to rupture of the oxide scale which is more dominant at 800–900°C [88]. Rupturing results in more diffusion of oxygen through the short-circuiting channel of the disintegrated surface. HfO_2 has anionic vacancies which trigger the oxygen diffusion. Also, HfO_2 has increased stability, elevated point defect generation enthalpy results in minimal defects [89]. However, a drop in oxidation with the increase in temperature to 1000°C has been reported due to the Hf recrystallization (range: 700–1400°C) which removes the residual stresses and enhances the plasticity [88]. The literature shows that oxidation of Hf at 1atm O_2 pressure and 1200°C is strictly linear till 30 min followed by a reduction in weight gain [90].

1.4 APPLICATIONS

Refractory metals have the ability to withstand high temperatures and therefore effectively are used in elevated temperature applications. However, with suitable modification of some of the refractory metals' properties at ambient temperatures, the application regime can be suitably extended. The applications of different refractory metals mentioned above are discussed in the following sections. The refractory metals are used pure or as a base element in an alloy, depending on the desired properties for an application.

1.4.1 W and W Alloys

W is the preferred metal for high temperature defense (kinetic energy penetrator), radiation shielding/plasma-facing material in nuclear reactors and rocket nozzle applications. W's application in a divertor is a potential approach. The major requirements of monoblock material W (Figure 1.1(a, b)) for a divertor are a superior elevated temperature creep (extended creep life), tensile, fatigue resistance, and inhibition of recrystallization and ductility [91]. The major requirements for the applicability of W for divertor application is the high purity of the starting material (≥ 99.94%), hardness ≥ 410 HV30, and finer grain (ASTM E112, Number 3) with elongated morphology [91]. High density and melting point (due to contact with large heat content on the surface) are also added advantages for W in such an application.

A high melting point and density are required for kinetic energy penetrators. Replacing depleted uranium (DU) due to environmental issues with W is an effective strategy with respect to penetrator application. Elevated kinetic energy along with a large length-to-diameter ratio provides an efficient ballistic performance. According to a report, the required properties for penetrator material is 6–8% ductility (minimum), fracture toughness of at least 33 MPa$\sqrt{m}$, increased strain rate of $10^6 s^{-1}$, and superior hydrostatic pressure (2–6 GPa) [92, 93] Therefore, the high density of W provides increased impact energy, the elevated strength value required for higher heat generation with respect to a particular strain, and restriction of precipitation of intermetallics which reduces the interfacial bonding [94]. Figure 1.2 shows a schematic representation of a kinetic energy penetrator. The fins are provided for a stabilization effect.

The material used for a rocket nozzle in elevated temperatures and pressure condition requires a superior melting point, thermal shock resistance, minimum thermal expansion coefficient and minimum erosion rate to achieve the desired propulsion [95]. Increased erosion increases the effective area which becomes problematic when it exceeds 5% and disrupts the nozzle expansion ratio and thrust behavior [96]. A high melting point and high hardness of W (influence to minimize wear) are useful properties in this context.

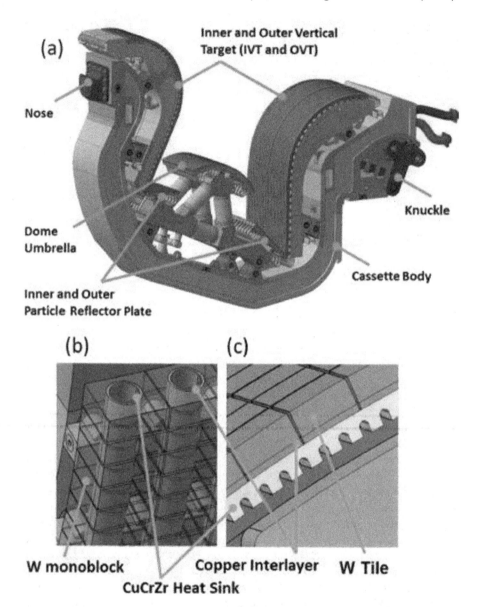

FIGURE 1.1 (a) Schematic view of ITER divertor consisting of inner and outer vertical targets, dome umbrella, dome particle reflector plates and cassette body; (b) monoblock geometry at the inner and outer vertical targets; (c) flat tile geometry at the dome umbrellas and particle reflector plates [91]. (Copyright (2016), with permission from Elsevier.)

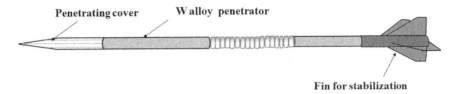

FIGURE 1.2 Schematic of kinetic energy penetrator: usage of tungsten. (Adapted from Wikipedia, "Kinetic energy penetrator.")

Heating elements in an elevated temperature furnace are also made of W sheet, wire and mesh, owing to its superior electrical conductivity. Apart from these, heating elements, the charge carrier, susceptors (for high frequency furnaces), and thermocouples are also made of W.

Due to the high density and elastic modulus of W, its usage in aviation and aerospace counterweights is preferred. W and W alloy aid in dampening vibration and enhance the durability of the components [97]. Electro-discharge machining electrodes, resistance welding electrodes, X-ray tubes anodes, filament for incandescent bulbs, scanning and transmission electron microscopes (due to low vapor pressure and high melting point) are some important applications of W and W-based alloys [97].

1.4.2 Mo and Mo Alloys

Mo is a widely accepted refractory metal used in a variety of high temperature applications. Its elevated temperature strength, stiffness, dimensional precision, elevated melting temperature, and low vapor pressure lead to its application in electronic tubes [33]. Due to good agreement with several compositions of molten glass, Mo is used as glass melting electrodes. The major advantages of Mo in this context are excellent current density, resistance to thermal shock, and high strength [98]. Shields in furnaces fabricated from Mo increase the sustainability of the life of the furnace. The application of Mo in a vacuum furnace hot zone is quite effective. The vapor pressure of Mo increases with the increase in temperature, however, the vapor pressure active in a vacuum furnace is higher than Mo, therefore, the sublimation tendency of pure Mo used as a tool in a vacuum furnace is suppressed. A comparison with W, which offers a poor fabricability issue compared to Mo, results in Mo's application in furnace components above 1500°C [98]. Mo possesses enhanced (50% higher) ability of heat conduction compared to steel and nickel alloys and excellent electrical conductivity, which result in its applicability as a heat sink [99]. Mo is also useful for coatings in piston rings and bearing shafts through powder and wire deposition technology.

X-ray detectors made from Mo sheet in the form of foil are efficient due to the high elastic modulus and excellent dimensional precision. The detector also protects from unwanted radiation and improves the efficiency of the required signal detection. Other exciting applications of Mo include flat panel displays (LCD) and solar cells due to: (1) appreciable electrical conductivity; (2) the similarity in the coefficient of thermal expansion, which relieves stress and further breaking; and (3) the tendency to form links between substrates [98].

Incandescent lamps use Mo as a support for the Tungsten filament. To adapt the stress generation when the lamp is heating up, sheets of Mo are used in suitable form for feedthrough [98]. Heat sinks made from Mo for power semiconductors, X-ray tube products for CT scans, bimetal thermocouples, airframes, cathodes in radar devices, and rockets are a few typical other applications of Mo and Mo alloys [98, 99].

1.4.3 Nb and Nb Alloys

The application of Nb can be categorized as:

1. steels (microalloyed);
2. steels (stainless and heat-resisting);
3. superalloys;
4. rest of applications (medical, nuclear).

So, the major contribution of Nb is in the steel industry as an alloying addition. In microalloyed steel or high strength low alloy steels (HSLA), Nb is added <0.1% content. Nb addition in steel forms Nb carbides followed by a drop in the solubility of carbon in the γ (austenite) and ferrite (α) phases. The second phase precipitate forms due to the strong affinity of Nb with C which results in a strengthening effect in steel. In tool steel, high temperature behavior is improved by the addition of Nb [42]. Nb-based steels are also preferred in the automotive industry due to their appreciable workability. Structural steel also uses Nb to enhance its strength.

The excellent gas absorption capacity of Nb facilitates its application in electron tubes operating under high vacuum. Nb in its pure form exhibits superconductivity (type II), an essential magnetic field and, therefore, Nb-based alloys are applied in superconducting coils [100]. Nb-based superalloys are an effective substitute for Ni-based superalloys for turbine applications due to the high melting point and low density. The operating temperature range of Nb-based alloys used as superalloys is higher compared to Ni-based superalloys. The thermal expansion coefficient and Young's modulus of Nb is almost 50% less than Ni in this context. Tanaka et al. have developed Nb-16Si-5Mo-15W-5Hf-5C (in atm%) alloy for gas turbine application and the alloy exhibits a creep rupture life greater than 100 h (<150 MPa, 1500°C condition), with specific strength: 50 MPa/Mg/m^3 (1500°C) [101].

Nb-1Zr alloy is preferred in sodium vapor lamp applications owing to its workability, weldability and durability in such applications. Nb-10Hf-1Ti alloy, commercially known as C-103 alloy, is employed in the aerospace industry within the temperature range of 1100–1500°C due its ability to retain strength. Valves used in high temperature, thrust cones, or nozzles are the use application areas of Nb alloys. Nb-based alloys are also useful in heat exchangers (due to chemically inert, good thermal conductivity), elevated temperature furnace parts, and reaction vessel applications [102, 103]. Nb can replace conventionally used Teflon sections in a heat exchanger in terms of cost factors [103].

Another important usage of Nb is in first reactor material in the nuclear sector. The high melting point, strength property, chemical resistance, and reduced neutron absorption of Nb are applicable as reactor material [103].

1.4.4 Re and Re Alloys

The cost and availability factor restrict the extensive application of Re. Strength at elevated temperature, insignificant permeability of gas (H_2) of Re are useful in solar rocket applications. Re coating is possible only in minimum O_2 concentration areas. Foils for identification of surplus oxygen in gas turbine engines and electrical vacuum components are made of Re. Electrical contact made from Re is effective compared to W due to noticeable electrical conductivity, resistance to wear and arc erosion [104–106]. Thermoionic and thermoelectronic emitters for mass spectroscopy are fabricated from Re in the form of ribbons or wire, owing to its high melting temperature, work function and the property which leads to efficient ionization, electrical resistance and preservation of flow property [107, 108]. Re thruster nozzles are subjected to a > 1,00000 fatigue cycle with an increase from room temperature to 2200°C. However, the thermal fatigue does not contribute to a breakdown of components as Re exhibits excellent creep strength at high temperatures [109]. Re developed from the CVD process has been acceptable as tubing used in thermocouple sheaths. Re combustion chambers fabricated by the powder metallurgical (PM) method have decreased production cost and time.

1.4.5 Ta and Ta Alloys

Ta usage is almost 66% in the electronics industry, 22% in the cutting tools industry, and as an alloy addition to superalloys to enhance high temperature strength, which accounts for 6%, 3% in chemical sectors (valves, heat exchangers, heaters), 1% in the medical industry, and 2% in the military sector (in kinetic energy projectiles) [110]. Several alloys of Ta have been developed to withstand the high temperature environment specifically for rockets and airframes. The developed Ta-10W, Ta-8W-2Hf, Ta-10W-2.5Hf-0.01C alloy shows a significant improvement in tensile strength and stress against rupture, compared to pure Ta. Ta-8W-2Hf alloy has been employed in space nuclear applications due to its strength, workability and weldability factors [110]. The ASTM F560 specification indicates the use of Ta as surgical implants. This is ascribed to its ductility and corrosion resistance and the biocompatibility of Ta. Although Ta possesses adequate biocompatibility, its elevated Young's modulus compared to bone results in a stress-shielding effect, which restricts the bio-implant usage of Ta. More specifically Ta is a potential refractory metal for porous implants [111]. Due to the higher melting point of Ta, it is quite challenging to fabricate by the casting method and it is directed to the development of Ta thin film metallic glass. Reports suggest that $Ta_{75}Ti_{10}Zr_8Si_7$ alloy exhibits lower Young's modulus, higher hardness (15.5 ± 0.3 GPa), less depth of scratch (179.1 ± 8.6 nm) compared to Ta (hardness: 12.5 ± 0.4 GPa, depth of scratch: 210.2 ± 3.4 nm) [112]. The increased hardness and reduced Young's modulus are a prerequisite for bio-implants. Ta can also endure in high temperature applications, such as nuclear reactors. The problem of radiation-based void formation in Ta is suppressed by the addition of W and overall protection against radiation is enhanced [113].

1.4.6 Ir AND Ir ALLOYS

The use of Ir in the chemical sectors in the form of electrodes is accredited to its enhanced resistance to corrosion, electronic conductivity and electrochemical steadiness [114]. Ir can effectively be electrodeposited on Ir wire and enhances the applicability of Ir electrodes [114]. Ir is also used in high temperature crucibles, electrical contacts, and added to improve the hardening effect of platinum, and compass bearings when it forms an alloy with Os [115]. Nanostructured Ir has significant importance in catalytic action. The enhancement is related to the high surface area of nanostructured Ir. The catalytic action of Ir is useful in the generation of the H_2 reaction in fuel cells [116]. Ir is also effective as coatings at high temperature as Ir exhibits enhanced creep resistance, in agreement in the coefficient of thermal expansion with all other refractory metals [117, 118]. A Re combustion compartment is also lined with Ir [119]. Due to the low oxidation rate of Ir at elevated temperatures, compared to W, Mo, Nb, Re, Ta, and its superior hardness, Ir coatings are efficient in retaining wear resistance. The literature also presents the application of Ir coatings on Re thruster chambers for satellite systems and the report also indicates appreciable durability of the coating in an oxidation temperature regime [120, 121].

1.4.7 Ru AND Ru ALLOYS

Ru belongs to the platinum group and due to its impact on conductivity, catalyst activity, and its mechanical property such as hardness, it finds several applications, including chip resistors and electrical contacts, in chemical sectors, and in solar cells (as a dye sensitizer). Ru-based compounds, specially RuO_2 are used as supercapacitors due to their elevated electrical conductivity value. Ru is formed during the fission reaction of UO_2 and MOX as fuel in nuclear plants and spent nuclear fuel also contains Ru, which is difficult to eliminate. Ru in oxidizing environments leads to the formation of RuO_2 and RuO_4 (volatile oxide), which leads to accidents in nuclear plants [122]. However, isotopes of Ru (^{97}Ru, ^{106}Ru) have significant importance in the treatment of cancer. Less toxic, fewer side effects afterwards are some of the contribution of Ru compounds in cancer treatment. Among the several states of oxidation of Ru (+II, +III, +IV), Ru (II) has enhanced impact in destroying tumor cells [123]. The major characteristics of Ru leading to use in medical areas are the ligand exchange rate, the available oxidation state variety, or imitating Fe to attach some biological molecules [124]. Antibiotics, antimalarial, immunosuppressant medicines are also Ru-based [124]. Ru is also used as a hardening for Pd and Pt alloys. Voltage regulators and thermostats use Pt-12%Ru base alloys and Ru also acts as an additive in Os alloys [125].

1.4.8 Os AND Os ALLOYS

Os or Os alloys have quite limited applications and mostly are used in the chemical and medical sectors, followed by the electrical industry in commercial products

Refractory Metals: Fundamentals   15

or process-related aspects. Needles, tips of fountain pen nibs, electrical contacts, bearings, phonograph needles are manufactured from Os or Os-based alloys. Heating filament is fabricated from pure Os. Incandescent lamps also use Os-W alloys (3–30 wt.% Os) as filaments, owing to high temperature stability (up to 2200°C) [126]. The alloy of Os-Ir is used as coatings owing to its hardness and wear resistance [127]. Os and Os oxides are also employed in the ceramic and glass industries. Osmium tetraoxide (OsO_4) finds useful application in microscopy stain and acts as a catalyst in developing certain drugs [128]. The mechanical properties (hardness) of Pt can be enhanced by the addition of Os. Reports point to the application of Os/W alloy in a dispenser cathode with lower work function than B-class emitters [129].

1.4.9 Hf AND Hf ALLOYS

Hf is used to manufacture control rods and emitters for the nuclear sector because of certain required properties, such as its neutron absorption capacity, elevated thermal conductivity, resistance to corrosion, and steadiness in dimension during radiation [130]. The corrosion resistance of Hf is superior to Zircaloy and almost 20 times higher than Ag-In-Cd. Reports indicate that the variation in diameter of Hf in a 15 years lifecycle is only 0.2% and the variation is non-significant with respect to design allowance [131]. Ni, Ta, Nb superalloys use considerable amounts of Hf as an alloying addition to endure high temperatures. Biocompatibility and the corrosion resistance of Hf alloys are effective in medical implants. Al-Mg-Sc alloys incorporate Hf to enhance strength for aerospace applications. SiO_2 is replaced by HfO_2 in manufacturing integrated circuits (ICs). Hf has strong bonding capacity with H_2 to form hydrides which have potential application in hydrogen storage [132–136]. Hf is also used for optical hydrogen sensing which is much more effective than Pd. The application of HfO_2 in electronics and the optoelectronics industry is sublime di-electric constant (almost 25), refractory index, less phonon frequency [137–139]. Golosov et al. have studied the effect of thickness of HfO_2 on the dielectric property [140]. The study reports that the dielectric constant drops with a reduction in the thickness of the film, and amorphous structure, elevated k-dielectrics are requirements for field effect transistors. Gas-filled and incandescent lamps and plasma welding torches are other important uses of Hf [132].

1.5 MAJOR LIMITATIONS WITH RESPECT TO APPLICATIONS

The refractory metals listed above are candidates for several high temperature applications. High melting point, low thermal expansion coefficient, high strength, toughness at elevated temperature properties are advantageous in this context. However, the above properties are material-specific and there are several limiting factors which restrict their applications. This section describes the major obstacles in the materials' properties which limits their effectiveness.

Although W has considerable significance in the nuclear industry, defense and space applications, some of its properties are not amenable to durability at high temperature. The rate of oxidation of W increases at elevated temperatures due to porous WO_3 formation. The presence of porosity and the volatility of oxides seriously degrade the endurance of material when exposed to increased temperature. Figure 1.3 shows that the weight change of pure W during oxidation increases with the increase in temperature from 800°C to 1000°C. The partial pressure of the formed WO_3 oxide also increases with temperature [141]. Due to increased WO_3 formation by consumption of base W followed by its sublimation, the penetration performance in a kinetic energy penetrator is reduced. Similarly, the high temperature oxidation of W in a rocket nozzle leads to rupture of the nozzle component. The volatility of WO_3 can be suppressed by applying a dense, adherent coating to the W component. The ductile-brittle transition temperature (DBTT) of W ranges between 200–500°C [16], therefore, below 200°C, W shows brittle behavior which restricts the room temperature applications. The higher DBTT of W can be counteracted by restricting grain coarsening and enhancing the recrystallization temperature. The limitation in the properties of W leads to the development of W-based alloys, such as W-Ni, W-Re [142, 143] which impart the plastic flow properties at room temperature. The literature shows considerable enhancement (92%) of room temperature fracture toughness of W by Ni doping [142].

One of the major problems with Mo is its extreme volatility in a high temperature environment which deteriorates its applicability as a heating element in furnaces. The DBTT of Mo is below room temperature (-23°C) [16], therefore, it is workable at elevated temperatures. A report indicates that room temperature brittleness leads to working of pure Mo at high temperature. The brittleness at room temperature causes the failure of Mo core heat pipes during conveying [144]. In a space nuclear power system, the acceptable working temperature is less than the recrystallization temperature of Mo, as, in the process of recrystallization, the strength drops. However, the creep resistance during high temperature working is very important. The coarser grains in the recrystallized structure offer better creep resistance. The maximum temperature limit of irradiation-based embrittlement is slightly higher than 1000 K (727°C) for Mo, whereas for W it is 1200 K (927°C) [145]. So, for Mo below the above-mentioned temperature, alloying is needed to improve the plastic flow properties for structural applications. Recrystallization of Mo leads to enhanced DBTT and therefore brittleness along with segregation of impurities, such as O, P, N, etc. at the grain boundary, which increases the propensity of fracture along the boundary and reduces the grain boundary strength [146].

Nb is ductile at room temperature as the DBTT of Nb is -126°C [15]. Nb has strong solubility for O, H, N, C. Oxidation of Nb initiates at 200°C, but oxide sublimation occurs beyond 1370°C (1643 K) [99] which reduces the effectiveness as the structural material. The ductility of Nb is also decreased at elevated temperatures due to oxygen incorporation [99]. In addition, the ultimate tensile strength and modulus of elasticity of Nb decrease with an increase in temperature, further restricting its applicability. Although from the point of view of lowest density, it is acceptable for structural applications, however, its weldability, adequate strength, and oxidation counteract it. Reports also suggest that though the addition of Mo to Nb is justified to improve the strength at elevated temperatures, there is a deleterious effect on fabricability [102].

Although Re possesses superior mechanical properties and corrosion resistance and therefore is used in coatings, space power, the nuclear industry, and the aircraft industry, however, its disadvantages are: (1) poor oxidation resistance; (2) high cost

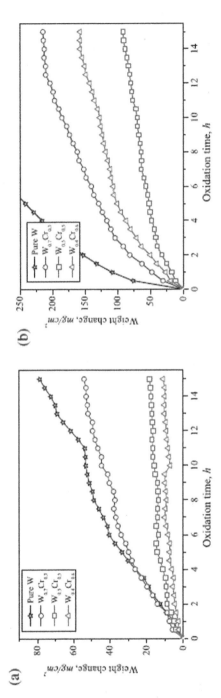

FIGURE 1.3 (a) Plots of (Δm/S) against (t) obtained from cyclic oxidation tests carried out in air on W–Cr alloys for duration of 15 h (30 cycles) at 800°C; (b) Plots of (Δm/S) against (t) obtained from cyclic oxidation tests carried out in air on W–Cr alloys for duration of 15 h (30 cycles) at 1000°C [141]. (Copyright (2013), with permission from Elsevier.)

of commercially available high purity powders or foil; and (3) increased time and energy factors during development [147]. For coating applications, chemical vapor deposition is used for Rhenium, and adding the reducing element Ni in high Re content leads to enhanced coating deposition efficiency by enhancing the current effectiveness [148]. Re is used as an alloying addition in W, Mo, Mg, etc. Mo-Re alloys with high Re content ($\geq$ 45%) form an intermetallic σ phase [149]. Carlen and Bryskin have reported that the Re content of 41–44.5% in Re-Mo alloys have no evidence of the σ phase and possess appreciable ductility at reduced temperature and elevated temperature strength [150]. Irradiation experiments on Mo-Re alloys at 800°C, 950°C and 1100°C exhibit enhanced ultimate tensile strength (> 1.5 GPa) with reduced ductility, but fracture along the grain boundary with respect to alloys subjected to aging and annealing [151]. Second-phase precipitation, segregation and transmutation have been identified as the major reasons for brittleness and it has been suggested to reduce the Re content to counteract the brittleness problem with respect to the application in fuel cladding and space reactors [151].

The elevated solubility of hydrogen in Ta is the major concern. The corrosion of Ta increases with the increase in temperature and associated brittleness, owing to the formation of hydrides which limit the application in a high temperature corrosive environment. The brittleness of Ta can be counteracted by the addition of a comparatively noble metal (platinum) [152]. Applications of Ta-based wet capacitors at low temperatures are difficult due to enhanced equivalent series resistance (ESR) and are related to a depreciation in capacitance and ripple current restrictions, in the application temperature zone [153]. The ESR is lowered and the capacitor damage is restricted with the development of MnO_2-Ta capacitors. The chances of a capacitor burning in an MnO_2-Ta capacitor is reduced by further development of polymer Ta capacitors which have comparatively less ESR and decreased temperature effect. However, the polymer capacitor has less operating voltage range and direct current leakage compared to wet Ta capacitors and MnO_2-Ta capacitors [153].

Although Ir has several applications as crucibles, rocket thrusters, coatings for electrode materials, and spark plugs, Ir coatings used in the space reactor undergo sublimation beyond 1100°C, which become very intense beyond 1227°C due to the formation of volatile IrO_3, which limits the durability of the components [154]. Elevated thermal expansion coefficient and less effective bonding with carbon also limit its coating applications, combined with the high cost and lack of accessibility [155]. Due to the poor fabricability of Ir, thermocouple application requires an Ir/Rh thermocouple for a high temperature oxidation atmosphere.

The application of Ru in the form of a catalyst with respect to the decrease in NO_x emission in the automobile sector is difficult, owing to the elevated temperature volatility [156]. The Ru oxides' behavior in the nuclear sector varies with the change in atmospheric condition (reducing or oxidizing). Significant Ru comes out during re-treatment of used fuel, due to the oxidation of the fuel [157, 158]. The literature evidence indicates that Ru is non-volatile in reducing the stream condition and in the primary heat transport system in a nuclear reactor, under such conditions Ru doesn't create any problem compared to the oxidation atmosphere which increases the danger [159]. Reports suggest that the vapor pressure of Ru at 2000 K with respect to 50% steam-hydrogen environment is almost 10^{-8} atm. and the hazard risk in a nuclear reactor is negligible [160]. Research has illustrated that Ru compounds contribute in cancer treatment [123], however, significant application of the compounds in medical sector is not evident. The high cost of RuO_2 and scaling-up of production limit its wide usage in capacitors, however, RuO_2 has effective charge storage efficiency [161].

Os in powder form produces OsO_4 during air heating and in an ambient atmosphere whereas bulk Os does not undergo oxidation. The other oxidation product of OsO_2 is stable. The OsO_4 compound is highly toxic and volatile. Though OsO_4 is toxic, reports have indicated no significant health problem or environmental concern by using Os in low concentrations. Due to its brittleness, Os is combined with Ir and Pt for applications in electrical contacts or tips of pen. The hardness of Os at elevated temperature (1200°C) is quite high compared to Ru and Ir, however, the room temperature ductility is quite a lot less as evident from % reduction (2–5%) compared to Ru (8–12%) and Ir (9–19%) through rolling [162]. The lower bulk to shear modulus ratio for Os (1.73) compared to platinum (4.52) [163] contributes to the brittleness and Os is alloyed with Pt to improve the workability for applications in pen tips.

Although Hf has appreciable strength, the major concern is the cost of its production. Hf shows corrosion resistance to several acids, including dilute HCl, H_2SO_4, and concentrated, dilute HNO_3. The presence of fluoride ions enhances the corrosion of Hf [164]. The solubility of hydrogen in Hf is quite a lot less and formation of hydrides leads to embrittlement and swelling of absorber material in nuclear reactor. Coated Hf is more prone to the problem compared to uncoated Hf, which leads to the formation of dense, abrasion-resistant HfO_2 film and reduces the generation of brittle hydrides [165]. During irradiation, Hf also deforms anisotropically and also minimizes the elongation of the claddings [165].

1.6 BASICS OF NANO-STRUCTURED MATERIALS: EXPERIMENTAL AND THERMODYNAMIC MODELING-BASED EVIDENCE

Nanostructured materials are defined as materials with dimensions less than 100 nm at least in one dimension. However, the final dimension of the material may vary with the synthesis routes and the corresponding process parameters adopted. The prime importance of nanostructured materials is based on: (1) lowering the processing temperature; and (2) improvement of mechanical properties. Selection of a suitable powder synthesis and sintering technique can lead to the development of near net shape products, i.e., with fewer or negligible finishing steps. Based on the dimension, chemical composition and accompanied fabrication techniques, nanomaterials can be classified into several categories [166]. Figure 1.4 shows the morphology of different varieties of nanostructured materials [167]. Due to the nanosize grain, the nanostructured materials show interesting mechanical, magnetic, and chemical properties. Zero-, one-, and two-dimensional nanostructured materials are represented as clusters, layered (lamellar) and filamentary respectively [166]. Generally available nanomaterials which are equiaxed in structure are defined as nanostructured crystallites (three-dimensional) which are synthesized by gas condensation or mechanical alloying [166]. The synthesis process which involves grinding the nature of the powder to be ductile or brittle decides the extent of the final refinement of the particle size.

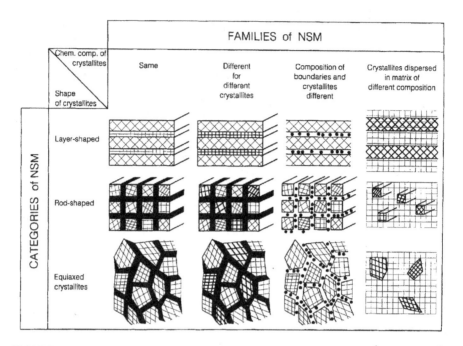

FIGURE 1.4 Classification for nanostructured materials according to their chemical composition and the dimensionality (shape) of the crystallites (structural elements) forming the nanostructured materials. (Adapted from [167]. Copyright (1995), with permission from Elsevier.)

Figure 1.5 shows the 2-D nanocrystalline material's structure with varied crystal orientation, and atoms in the crystals and atoms at the interface are displayed. The boundary core regions (open circle) possess decreased atomic density. There is variation in the interatomic distance of the boundary region [168]. Reduced energy regions and misfit areas (boundary areas) in the nanocrystalline materials develop the non-equilibrium structure. The boundaries in the nanocrystalline material are quite significant (around 6×10^{25} m^{-3}) with respect to the 10 nm grain size. The boundary-related atom fraction (volume) can be evaluated from [166]:

$$C = 3\Delta / d \qquad (1.3)$$

where Δ is average grain boundary thickness, d is average of grain diameter, the grains are supposed to be sphere, cubes. Therefore, with an increase in average grain diameter from 5 nm to 100 nm, the atom fraction in volume at the interface decreases from 50% to 3% respectively [166].

The 3-D nanostructured materials are investigated with respect to strength, workability and magnetic behavior, whereas 1-D nanostructured materials are intended with respect to electronic usage. Fabrication of nanocrystallites is effective to improve the solid solubility, which it is challenging to achieve using conventional melting-casting

Refractory Metals: Fundamentals

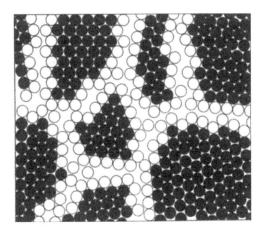

FIGURE 1.5 Atomic structure in two-dimensional nanocrystalline material. (Reprinted from [168]. Copyright (1989) with permission from Elsevier.)

Note: The structure was calculated by following the method provided in [169]. The atomic interaction was denoted by a Morse potential fitted to gold [169]. Centers of the "Crystals" atoms are presented in black. Open Circles are the atoms in the boundary core regions.

techniques. The nano-structured materials with high chemical reactivity exhibit a reduced consolidation temperature and result in enhanced strength. Nanomaterials also exhibit higher dangling bonds and result in reduced cohesive energy which is responsible for a reduction in the melting temperature and enhancement in the energy bandgap [170, 171]. A surface polarized bond in nano-materials causes a reduction in the dielectric constant. The cohesive energy of nanomaterials (E_{cn}) is presented according to the literature evidence as [172]:

$$E_{cn} = E_0 \left(1 - \frac{3N}{4n}\right) \quad (1.4)$$

where E_0 is the cohesive energy of bulk material, n is the number of total atoms in nanomaterial, N is the surface atoms number.

The melting temperature of the bulk nanomaterials (T_{mB}) can be presented as [173]:

$$T_{mB} = \frac{0.032}{k} E_0 \quad (1.5)$$

where k is the constant,

The melting temperature of the nanomaterials (T_{mN}) and the bulk nanomaterials (T_{mB}) are related as:

$$T_{mN} = T_{mB} \left(1 - \frac{3N}{4n}\right) \quad (1.6)$$

The melting temperature of the nanomaterials and the bulk size is also related to the shape factor (α_{shape}) as [174, 175]:

$$\frac{T_{mN}}{T_{mB}} = 1 - \frac{\alpha_{shape}}{D} \qquad (1.7)$$

where D is the nanoparticle size. The shape parameter (factor) depends on the morphology of the nanoparticles. The shape parameters for spherical nanoparticles, regular tetrahedral nanoparticles, regular octahedral nanoparticles, cylindrical nanowires, regular triangular cross-section nanowires, regular hexagonal cross-section nanowires are 1, 1.49, 1.18, 1, 1.286, 1.05 respectively [176, 177].

The energy bandgap (E_g) is the difference between the Gibbs free energy among the conduction and valence energy bands and can be represented as [178]:

$$E_g = \Delta H_{cv} - T\Delta S_{cv} \qquad (1.8)$$

where ΔH_{cv} is the difference of enthalpy among conduction and valence energy bands, ΔS_{cv} is the difference of entropy among conduction and valence energy bands.

The energy bandgap enhances with a reduction in the size of the nanomaterials. The energy bandgap of the nanoparticles and bulk nanomaterial is represented as E_{gN} and E_{gB} respectively, which is related to the melting point as [179]:

$$\frac{E_{gN}}{E_{gB}} = 2 - \frac{T_{mN}}{T_{mB}} \qquad (1.9)$$

The energy bandgap increases with a reduction in the melting temperature of the nanomaterials and a reduction in size owing to [180]:

1. enhancement of the surface atoms;
2. the quantam captivity of electrons, holes and energy for transition enhancement from the valence band's top level to the conduction band bottom level.

The literature on the variation of the energy bandgap and size (nm) shows that the energy bandgap increases with the decrease in the size [170, 181–185]. The viability of Equation (1.9) is presented in Figure 1.6 which shows that the quantam confinement effect of CdS nanoparticles, where the size of the CdS nanoparticles are very small, can be related to the electron's wavelength. Eg and Eg, ∞ denote the energy bandgap related to the size-dependent and bulk melting temperature.

$$\frac{E_g}{E_{g,\infty}} = 1 + \frac{\alpha_{sphere}}{D} \alpha_{sphere} \qquad (1.10)$$

The high strength of the nanomaterials is due to the increased grain boundary area which inhibits the dislocation slip. The very fine grains in nanomaterials are unable to store the dislocation which minimizes their ductility. The strength-grain size

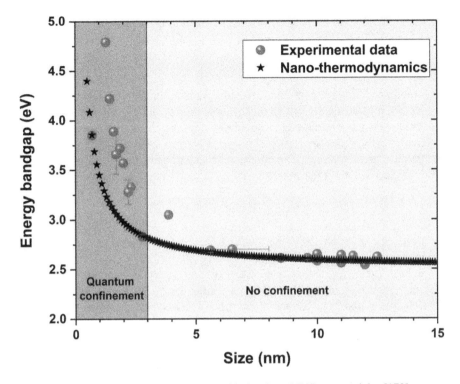

FIGURE 1.6 Variation of energy bandgap with the size of CdS nanoparticles [170]. Note: The experimental data is from[181–184]. The scaling law, $\frac{E_g}{E_{g,\infty}} = 1 + \frac{\alpha_{sphere}}{D}$ is applied to envisage the energy bandgap of CdS nanoparticles and shape parameter, $\alpha_{sphere} = 0.38$ nm [185].

relationship is for nanomaterials with a grain size higher than almost 10–30 nm. However, with a reduction in the grain size below the range, an inverse relationship exists, which is an Inverse Hall-Petch relationship, as the grains are too small so that dislocation formation in the grains is hindered. The structural materials require high strength and adequate toughness in applications. High temperature processing leads to improved ductility, however, with reduced strength. Metals with high DBTT, such as W, show appreciable ductility at elevated temperatures, however, this is lacking in room temperature applications due to brittleness. Coarser grains are efficient dislocation storage areas which lead to increased strain hardening and improved ductility. Several methods are suggested to improve the strength and ductility in nanostructured materials, such as the following [186]:

1. Generation of nano twins and increase in the strain hardening effect.
2. Increase in the density of stacking faults.
3. Precipitation hardening mechanism.

4. Development of heterogeneous lamella structures and metals with graded structures.
5. Incorporation of carbon nanotubes in the grain matrix of nanostructured materials.

To comprehend the strength-ductility enhancement in heterogeneous structures, the density of statistically stored dislocation (results of the interaction between dislocation-soft particles and dislocation-alloying constituents) (ρ_{ss}) and the density of geometrically necessary dislocations (ρ_{gn}) (results of dislocation-hard particle interactions) are important. The above dislocation densities are related to work hardening (σ_d) as [187, 188]:

$$\sigma_d = \alpha MGb\sqrt{\rho_{ss} + \rho_{gn}} \quad (1.11)$$

where α is the constant, M is the Taylor factor, G is the shear modulus, b is the Burgers vector.

The back stress contributed by geometrically necessary dislocations increases the yield strength and increased back stress hardening improves the ductility.

Twins generation by deformation contributes to enhanced strength and ductility in nanostructured materials. The twinning mechanism depends on the grain size and the variation of twinning stress with the grain size of HCP, BCC, FCC structures is presented in Figure 1.7 [186]. The tendency for deformation twins to be generated reduces with a reduction in the grain size due to the relaxation of the stress concentration, which arises from the rotation of the grains and grain boundary sliding. Improvement in strength and ductility by deformation twins is more effective in FCC

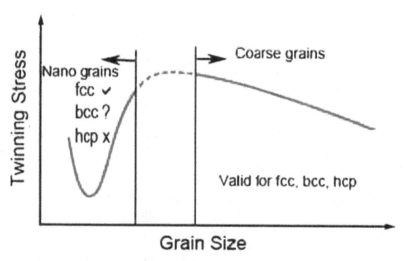

FIGURE 1.7 Schematics of the grain size effect on deformation twinning for FCC, BCC, and HCP metals. (Reprinted from [186]. Copyright (2018), with permission from Elsevier.)

nanocrystalline materials. The effect of grain size on twins generation in BCC metals is not properly comprehended [186].

Ductility of Mo around room temperature is also inadequate. The process of making nanostructured Mo alloys to achieve elevated yield strength at room temperature (above 800 MPa) and enriched elongation in tensile mode of around 40%, is reported in the literature [189]. The report suggests that doping of La_2O_3 oxide particles in the Mo matrix results in the presence of La_2O_3 particles in the Mo matrix. The La_2O_3 particles facilitate the formation and accumulation of dislocation in the grain matrix and enhance the ductility by restricting the crack at the grain interface.

The oxidation behavior of nanostructured materials depends on the effect of the grain boundary area. An increased grain boundary region in nanomaterials enhances the presence of several defects, such as vacancies or dislocations. The higher grain boundary area also increases the diffusivity and electrical resistivity [190]. Therefore, grain boundary areas are effective sites for the diffusion of oxygen for further oxide formation. If the nature of the oxide is porous with high sublimation rate, the grain boundary acts as an activator for elevated oxidation. However, due to the faster diffusion of oxygen ions through the grain boundary to the surface, metals which produce a comparatively protective layer show improvement in oxidation resistance. The grain boundary of nanostructured materials with poor oxidation resistance needs to be coated with a protective layer with improved cohesion with the substrate. Apart from the density of the oxide layer, the cohesion/adhesion is vital to inhibit spallation or blistering due to a repeated thermal cycle during oxidation. Suitable alloying and oxide dispersion can inhibit the crack propagation and can improve the oxidation resistance. More comprehensive details on the oxidation behavior of nanomaterials will be provided in Chapter 8.

1.7 INFERENCES

In the above sections the physical, mechanical, high temperature properties of refractory metals are discussed. The wide application areas of refractory metals and alloys are also illustrated. However, several refractory metals, such as Mo, Ta, Nb, and Ru, exhibit inadequate ultimate tensile strength, whereas W and Ir show slightly higher UTS compared to Mo, Ta, Nb, Ru at an elevated temperature (1000°C). The restricted percentage elongation of W and Mo at room temperature also impacts their workability. The reduced hardness with increase in temperature also contributes to a reduction in wear resistance of the refractory metals which has an immense impact on the high temperature applications, such as furnaces, rocket nozzles, etc. Improving the creep strength and creep rupture life of W and Mo is of major importance. Many refractory metals, such as W, Mo, Nb, Re, Ta, Ru, and Os, exhibit considerable oxidation compared to Ir and Hf at elevated temperatures. The oxidation products are porous and volatile which cause significant failure of components in structural applications. Enhanced strength with adequate ductility of refractory metals during high temperature and room temperature applications and enhanced high temperature oxidation resistance should be the major thrust areas for research.

REFERENCES

[1] J. L. Johnson, Sintering of refractory metals, in: Z. Z. Fang (Ed.), *Sintering of Advanced Materials: Fundamentals and Processes*, Woodhead Publishing Limited, Cambridge, (2010) 356–388.
[2] H. L. Skriver, Crystal structure from one-electron theory, *Phys. Rev. B*, 31 (1985) 1909–1923.
[3] J. H. Dai, W. Li, Y. Song, L. Vitos, Theoretical investigation of the phase stability and elastic properties of TiZrHfNb-based high entropy alloys, *Mater. Des.*, 182 (2019) 108033.
[4] J. R. L. Trasorras, T. A. Wolfe, W. Knabl, C. Venezia, R. Lemus, E. Lassner, W. D. Schubert, E. Lüderitz, H. U. Wolf, Tungsten, tungsten alloys, and tungsten compounds, in: *Ullmann's Encyclopedia of Industrial Chemistry*, Wiley-VCH Verlag GmbH & Co. Weinheim, (2016) 1–53.
[5] D. Pettifor, *Bonding and Structure of Molecules and Solids*, Clarendon Press, Oxford, (1995).
[6] V. Levitin, *Interatomic Bonding in Solids: Fundamentals, Simulation, and Applications*, John Wiley & Sons, Chichester, (2014).
[7] C. Kittel, *Introduction to Solid State Physics*, 5th edn. Wiley, Hoboken, NJ, (1976).
[8] Y. Giret, S. L. Daraszewicz, D. M. Duffy, A. L. Shluger, K. Tanimura, Nonthermal solid-to-solid phase transitions in tungsten, *Phys. Rev. B*, 90 (2014) 094103.
[9] C. T. Chan, D. Vanderbilt, S. G. Louie, J. R. Chelikowsky, Theoretical study of the cohesive and structural properties of Mo and W in BCC, FCC, and HCP structures, *Phys. Rev. B*, 33 (1986) 7941–7946.
[10] A. B. Belonoshko, S. I. Simak, A. E. Kochetov, B. Johansson, L. Burakovsky, D. L. Preston, High-pressure melting of molybdenum, *Phys. Rev. Lett.*, 92, (2004) 195701.
[11] A. B. Belonoshko, L. Burakovsky, S. P. Chen, B. Johansson, A. S. Mikhaylushkin, D. L. Preston, S. I. Simak, D. C. Swift, Molybdenum at high pressure and temperature: melting from another solid phase, *Phys. Rev. Lett.*, 100 (2008) 135701.
[12] W. F. Gale, T. C. Totemeier (Eds.), *Crystal Chemistry: Smithells Metals Reference Book*, 8th edn, Butterworth-Heinemann, Oxford, (2004) 61–62.
[13] P. Duwez, The allotropic transformation of hafnium, *J. Appl. Phys.*, 22 (1951) 11–74.
[14] J. D. Fast, The allotropic transformation of hafnium and a tentative equilibrium diagram of the system zirconium-hafnium, *J. Appl. Phys.*, 23 (1952) 350
[15] A. K. Suri, Refractory metals and alloys, in: R. Chidambaram, S. Banerjee (Eds.), *Materials Research: Current Scenario and Future Projections*, Allied Publishers: New Delhi, (2003) 275–300.
[16] J. W. Semmel, Oxidation behavior of refractory metals and alloys, in: M. Semahyshen, J. J. Harwood (Eds.), *Refractory Metals and Alloys*, Interscience, New York, (1961).
[17] S. Watanabe, S. Nogami, J. Reiser, M. Rieth, S. Sickinger, S. Baumgärtner, T. Miyazawa, A. Hasegawa, Tensile and impact properties of tungsten-rhenium alloy for plasma-facing components in fusion reactor, *Fusion Eng. Des.*, 148 (2019) 111323.
[18] O. Boser, The temperature dependence of the flow-stress of heavily-deformed doped tungsten, *J. Less-Common Met.*, 23 (1971) 427–435.
[19] X. Y. Ding, L. M. Luo, L. M Huang, G. N. Luo, X. Y. Zhu, J. G. Cheng, Y. C. Wu, Preparation of TiC/W core–shell structured powders by one-step activation and chemical reduction process, *J. Alloys Compd.*, 619 (2015) 704–708.
[20] S. Wurster, B. Gludovatz, R. Pippan, High temperature fracture experiments on tungsten-rhenium alloys, *Int. J. Refract. Met. Hard Mater.*, 28 (2010) 692–697.

[21] E. Gaganidze, D. Rupp, J. Aktaa, Fracture behaviour of polycrystalline tungsten, *J. Nucl. Mater.*, 446 (2014) 240–245.
[22] B. Gludovatz, S. Wurster, A. Hoffmann, R. Pippan, Fracture toughness of polycrystalline tungsten alloys, *Int. J. Refract. Met. Hard Mater.*, 28 (2010) 674–678.
[23] B. C. Allen, D. J. Maykuth, R. I. Jafee, The recrystallization and ductile- brittle transition behaviour of tungsten: Effect of impurities on polycrystals prepared from single crystals, *J. Inst. Met.*, 90 (1961) 120–128.
[24] S. Robinson, O. Sherby, Mechanical behavior of polycrystalline tungsten at elevated temperature, *Acta Metall.*, 17 (1969) 109–125.
[25] C. Barrett, O. Sherby, Influence of stacking fault energy on high temperature creep of pure metals, *Trans. Metall. Soc. AIME*, 233 (1965) 1116–1119.
[26] W. D. Klopp, P. L. Raffo, Effects of purity and structure on recrystallization, grain growth, ductility, tensile strength and creep properties of arc melted tungsten, National Aeronautics and Space Administration, (NASA TN D-2503), (1964).
[27] W. D. Klopp, W. R. Witzke, Mechanical properties and recrystallization behavior of electron-beam-melted tungsten compared with arc-melted tungsten, National Aeronautics and Space Administration, (NASA TN D-3232), (1966).
[28] G. King, H. Sell, The effect of thoria on the elevated-temperature tensile properties of recrystallized high purity tungsten, *Trans. Metall. Soc. AIME*, 82 (5) (1965) 1104–1113.
[29] J. Conway, P. Flagella, Physical and mechanical properties of reactor materials, Seventh Annual Report, AEC Fuels and Materials Development Program, (1968).
[30] B. Harris, E. Ellison, Creep and tensile properties of heavily drawn tungsten wire, *Trans. of ASM*, 59 (1966) 744–754.
[31] W. Green, Short time creep rupture behavior of tungsten at 2250 to 2800°C, *Trans. Metall. Soc. AIME*, 215 (1959) 1058–1161.
[32] E. A. Gulbransen, Thermochemistry and the oxidation of refractory metals at high temperature, *Corrosion*, 26 (1970) 19–28.
[33] C. K. Gupta, *Extractive Metallurgy of Molybdenum*, Routledge, London, (2017).
[34] D. Sturm, M. Heilmaier, J. H. Schneibel, P. Jéhanno, B. Skrotzki, H. Saage, The influence of silicon on the strength and fracture toughness of molybdenum, *Mater. Sci. Eng. A*, 463 (2007) 107–114.
[35] A. Luo, J. J. Park, D. L. Jacobson, B. H. Tsao, M. L. Ramalingam, Creep behavior of molybdenum and a molybdenum-hafnium carbide alloy from 1600 to 2100 K, *Mater. Sci. Eng. A*, 177 (1994) 89–94.
[36] G. S. Cho, G. B. Ahn, K. H. Choe, Creep microstructures and creep behaviors of pure molybdenum sheet at 0.7 Tm, *Int. J. Refract Met. Hard Mater.*, 60 (2016) 52–57.
[37] E. A. Gulbransen, K. F. Andrew, F. A. Brassart, Oxidation of molybdenum 550° to 1700°C, *J. Electrochem. Soc.*, 110 (1963) 952–959.
[38] M. Simnad, A. Spilners, Kinetics and mechanism of the oxidation of molybdenum, *JOM*, 203 (1955) 1011–1016.
[39] C. English, The physical, mechanical and irradiation behaviour of niobium and niobium base alloys, in: H. Stuart (Ed.), *Proc. Int. Symp. on Niobium* 81, The Met. Soc. American Institute of Mining, Metallurgical and Petroleum Engineers (AIME), Warrendale, PA, (1984) 239.
[40] R. P. Elliot, *Constitution of Binary Alloys, 1st Supplement*, McGraw-Hill, New York, (1965) p. 261.
[41] R. T. Begley, D. L. Harrod, R. E Gold, High temperature creep and fracture behaviour of the refractory metals, in: I. Machin, R. T. Begley, E. D. Weisert (Eds.), *Refractory Metals Alloys: Metallurgy and Technology*, Plenum Press, New York, (1968) 41.

[42] C. K. Gupta, A. K. Suri, *Extractive Metallurgy of Niobium*, CRC Press, Boca Raton, FL, (1993).

[43] C. Gandhi, M. F. Ashby, Fracture-mechanism maps for materials which cleave: F.C.C, B.C.C. and H.C.P. metals and ceramics, *Acta Metall.*, 27 (1979) 1565–1602.

[44] J. P. Guha, Studies on niobium oxides and polymorphism of niobium pentoxide, *Trans. Indian Ceram. Soc.*, 28 (4) (1969) 97–101.

[45] O. Kubaschewski, B. E. Hopkins, Oxidation mechanisms of niobium, tantalum, molybdenum and tungsten, *J. Less-Common Met.*, 2 (1960) 172–180.

[46] N. N. Greenwood, A. Earnshaw, Manganese, technetium and rhenium, in: *Chemistry of the Elements*, 2nd edn, Butterworth-Heinemann, Oxford, (1997) 1040–1069.

[47] J. R. Davis, *ASM Specialty Handbook: Heat-Resistant Materials*, ASM International, Materials Park, OH, (1997).

[48] W. R. Witzke, P. L. Raffo, Creep behavior of electron-beam-melted rhenium, *Metall. Trans.*, 2 (1971) 2533–2539.

[49] T. C. Chou, A. Joshi, C. M. Packer, Oxidation behavior of rhenium at high temperatures, *Scr. Metall. et Mater.*, 28 (12) (1993) 1565–1570.

[50] C. L. Briant, M. K. Banerjee, *Refractory Metals and Alloys: Reference Module in Materials Science and Materials Engineering*, Elsevier, Oxford, (2016).

[51] R. B. Kotelnikov, S. N. Bashlykov, Z. G. Galiakbarov, A. I. Kashtanov, Osobo tugoplavkie elementy i soedineniya [Extra refractory elements and compounds]. *Metallurgiya*, Moscow (1968) [in Russian].

[52] S. M. Cardonne, P. Kumar, C. A. Michaluk, H. D. Schwartz, Tantalum and its alloys, *Int. J. Refract, Met. Hard Mater.*, 13 (1995) 187–194.

[53] J. Vanaja, K. Laha, M. D. Mathew, T. Jayakumar, E. R. Kumar, Effects of Tungsten and tantalum on creep deformation and rupture properties of reduced activation ferritic-martensitic steel, *Procedia Eng.*, 55 (2013) 271–276.

[54] V. B. Voitovich, V. A. Lavrenko, V. M. Adejev, E. I. Golovko, High-temperature oxidation of tantalum of different purity, *Oxid. Met.*, 43 (1995) 509–526.

[55] P. Kofstad, *High-Temperature Oxidation of Metals*, Wiley, New York, (1966).

[56] P. Kofstad, The oxidation behavior of tantalum at 700°–1000°C, *J. Electrochem. Soc.*, 110 (1963) 491.

[57] N. Birks, G. H. Meier, *Introduction to High Temperature Oxidation of Metals*, Edward Arnold, London, (1983).

[58] A. V. Ermakov, S. S. Naboichenko, Iridium: Production, consumption, and prospects. *Russ. J. Non-ferrous Metals*, 53 (2012) 292–301.

[59] A. G. Degussa: *Edelmetall-Taschenbuch*, 2nd edn. Huthig, Heidelberg, (1995).

[60] R. Weiland, D. F. Lupton, B. Fischer, J. Merker, C. Scheckenbach, J. Witte, High-temperature mechanical properties of the platinum group metals, *Platinum Metals Rev.*, 50 (4) (2006) 158–170.

[61] R. L. Mehan, E. C. Duderstadt, E. H. Sayell, Creep-rupture behavior of iridium at elevated temperatures, *Metall. Trans. A*, 6 (1975) 885–889.

[62] J. C. Chaston, Reactions of oxygen with the platinum metals. *Platinum Metals Rev.*, 9 (1965) 51–56.

[63] R. T. Wimber, H. G. Kraus, Oxidation of iridium, *Metall. Trans.*, 5 (1974) 1565–1571.

[64] Doduco, *Datenbuch, Handbuch für Techniker*, 2nd edn., Doduco, Pforzheim (1977).

[65] L. Zhu, S. Bai, H. Zhang, Y. Ye, W. Gao, Long-term high-temperature oxidation of iridium coated rhenium by electrical resistance heating method, *Int. J. Refract Met. Hard Mater.*, 44 (2014) 42–48.

[66] A. G. Degussa: *Edelmetall-Taschenbuch*, Degussa, Frankfurt, (1967).

[67] W. Song, et al., Effect of ruthenium on microstructure and high-temperature creep properties of fourth generation Ni-based single-crystal superalloys, *Mater. Sci. Eng. A*, 772 (2020) 138646.

[68] Z. Guoqi, T. Sugui, S. Delong, T. Ning, Y. Huajin, W. Guangyan, L. Lirong, Influence of Ru on creep behaviour and concentration distribution of Re-containing Ni-based single crystal superalloy at high temperature, *Mater. Res. Express*, 7 (2020) 066507.

[69] J. C. Chaston, The oxidation of the platinum metals, *Platinum Metals Rev.*, 19 (4) (1975) 135–140.

[70] Z. Hózer, L. Matus, P. Windberg, Codex and ruset air oxidation experiments, 5th Technical Seminar Phebus FP Programme, Aix-en-Provence, France, June 24–26, (2003).

[71] C. Madic, C. Mun, L. Cantrel. Review of literature on ruthenium behaviour in nuclear power plant severe accidents. *Nuclear Technology*, 156 (3) (2017) 332–346.

[72] W. Martienssen, The elements, in: W. Martienssen, H. Warlimont (Eds.), *Springer Handbook of Condensed Matter and Materials Data*, Springer, Heidelberg, (2005) 45–158.

[73] R. B. Ross, *Metallic Materials Specification Handbook*, 4th edn, vol. 1, Springer US, New York, (1992).

[74] R. T. Grimley, R. P. Burns, M. G. Inghram, Mass-spectrometric study of the osmium-oxygen system, *J. Chem. Phys.*, 33 (I) (1960) 308–309.

[75] C. A. Krier, R. I. Jaffee, Oxidation of the platinum-group metals, *J. Less-Common Metals*, 5 (1963) 411–431.

[76] W. L. Phillips, Jr., Oxidation of the platinum metals in air, *Trans. Am. Soc. Met.*, 57 (1964) 33–37.

[77] R. Tricot, The metallurgy and functional properties of hafnium, *J. Nucl. Mater.*, 189 (1992) 277–288.

[78] J. H. Schemel, Manual on zirconium and hafnium, *ASTM-STP*, 639 (1977).

[79] J. Detours, Internal report, SRMA-CEA Saclay, (1986).

[80] M. Armand, Zirconium et Hafnium. Métallurgie et applications, *Bulletin du Cercle d'Etudes des Métaux XII*, (1972) 157.

[81] R. Tricot, Le zirconium dans le génie chimique. Caractères principaux de la construction chaudronnée, *Matériaux et Techniques*, April–May (1989) 1.

[82] D. Charquet, Propriétés du zirconium et du hafnium, *Techniques de l'ingénieur, Traité de métallurgie M*, 560 (October) (1985).

[83] H. Frost, M. Ashby: *Deformation-Mechanism Maps*, Pergamon Press, Oxford, (1982).

[84] H. Y. Zhu, X. F. He, Z. R. Liu, The preferred slip plane of nuclear material of Hafnium: A first-principles study, *Comput. Mater. Sci.*, 157 (2019) 25–30.

[85] D. R. McClintock, G. A. Tracy, Physical and mechanical properties of hafnium, *WAPD-SFR-Met-653*, January 7, (1958).

[86] R. F. Domagala, R. Ruh, The hafnium-oxygen system, *Trans. Am. Soc. Metals*, 58 (1965) 164.

[87] W. W. Smeltzer, R. R. Haering, J. S. Kirkaldy, Oxidation of metals by short circuit and lattice diffusion of oxygen, *Acta Metall.*, 9 (1961) 880–885.

[88] V. B. Voitovich, V. A. Lavrenko, E. I. Golovko, V. M. Adejev, The effect of purity on the high-temperature oxidation of hafnium, *Oxid. Met.*, 42 (1994) 249–263.

[89] P. Kofstad, Effect of impurities on the defects in oxides and their relationship to oxidation of metal, *Corrosion*, 24 (1968) 379–388.

[90] P. Kofstad, S. Espevik, Kinetic study of high-temperature oxidation of hafnium, *J. Less-Common Metals*, 12 (1967) 382–394.

[91] T. Hirai, et al., Use of tungsten material for the ITER divertor, *Nucl. Mater. Energy*, 9 (2016) 616–622.

[92] W. E. Gurwell, in: A. Crowson, E. S. Chen (Eds.), *Tungsten and Tungsten Alloys – Recent Advances*, TMS, Warrendale, PA, (1991) 43–52.
[93] P. Woolsey, R. J. Dowding, K. J. Tauer, F. S. Hodi, Performance-property relationships in tungsten alloy penetrators, in A. Bose, R. J. Dowding (Eds.), *Tungsten and Tungsten Alloys – 1992*, Metal Powder Industries Federation, Princeton, NJ, (1993) 533–540.
[94] A. Upadhyaya, Processing strategy for consolidating tungsten heavy alloys for ordnance applications, *Mater. Chem. Phys.*, 67 (2001) 101–110.
[95] D. Y. Park, Y. J. Oh, Y. S. Kwon, S. T. Lim, S. J. Park, Development of non-eroding rocket nozzle throat for ultra-high temperature environment, *Int. J. Refract Met. Hard Mater.*, 42 (2014) 205–214.
[96] E. Y. Wong, Solid rocket nozzle design summary, AIAA Paper 68-665, paper presented at 4th AIAA Propulsion Joint Specialist Conference, Cleveland, OH, (1968).
[97] E. Lassner, W. D. Schubert, *Tungsten: Properties, Chemistry: Technology of the Element, Alloys, and Chemical Compounds*, Kluwer Academic/Plenum Publishers, New York, (1999).
[98] J. A. Shields, *Applications of Molybdenum Metals and its Alloys*, 2nd edn, International Molybdenum Association (IMOA), London, (2013).
[99] O. D. Neikov, S. S. Naboychenko, I. B. Murashova, N. A. Yefimov, Production of refractory metal powders, in: O. D. Neikov, S. S. Naboychenko, N. A. Yefimov (Eds.), *Handbook of Non-Ferrous Metal Powders*, 2nd edn, Elsevier, Oxford, (2019) 685–755.
[100] J. Eisenstein, Superconducting elements, *Rev. Mod. Phys.*, 26 (1954) 277–291.
[101] R. Tanaka, A. Kasama, M. Fujikura, I. Iwanaga, H. Tanaka, Y. Matsumura, Research and development of niobium-based superalloys for hot components of gas turbines, in *Proceedings of the International Gas Turbine Congress 2003*, (2003) 1–5.
[102] V. V. Satya Prasad, R. G. Baligidad, A. A. Gokhale, Niobium and other high temperature refractory metals for aerospace applications, in: N. Prasad, R. Wanhill (Eds.), *Aerospace Materials and Material Technologies*, Indian Institute of Metals Series, Springer, Singapore, (2017) 267–288.
[103] R. W. Balliett, M. Coscia, F. J. Hunkeler, Niobium and tantalum in materials selection. *JOM*, 38 (1986) 25–27.
[104] NASA, Rhenium-foil witness cylinders, *NASA Tech Briefs*, 42, (1992).
[105] E. M. Savitskii., M. A. Tylkina, paper presented at Electrochemical Society, Symposium on Rhenium, Chicago, (1960).
[106] C. T. Sims, *Metal Industry*, Battelle Memorial Institute, Columbus, OH, (1955) 1–4.
[107] J. J. Manura, *The Mass Spectrometer Source*, Scientific Instrument Services, New Jersey, (1990) 14–19.
[108] J. C. Carlen, B. D. Bryskin, Rhenium: A unique rare metal, *Mater. Manuf. Processes*, 9 (6) (1994) 1087–1104.
[109] J. R. Wooten, P. T. Lansaw, The enabling technology for long-life, high performance on-orbit and orbit-transfer propulsion systems: High temperature, oxidation-resistant thrust chambers, in *Proceedings of the 1989 JAANAF Propulsion Meeting*, Chemical Propulsion Information Agency, Laurel, MD, (1989).
[110] R. W. Buckman, New applications for tantalum and tantalum alloys, *JOM*, 52, (2000) 40–41.
[111] K. Vanmeensel, K. Lietaert, B. Vrancken, S. Dadbakhsh, X. Li, J. P. Kruth, P. Krakhmalev, I. Yadroitsev, J. V. Humbeeck, Additively manufactured metals for medical applications, in: J. Zhang, Y. G. Jung (Eds.), *Additive Manufacturing*, Butterworth-Heinemann, Oxford, (2018) 261–309.

[112] J. J. Lai, Y. S. Lin, C. H. Chang, T. Y. Wei, J. C. Huang, Z. X. Liao, C. H. Lin, C. H. Chen, Promising Ta-Ti-Zr-Si metallic glass coating without cytotoxic elements for bioimplant applications, *Appl. Surf. Sci.*, 427 (2018) 485–495.

[113] I. Ipatova, P. T. Wady, S. M. Shubeita, C. Barcellini, A. Impagnatiello, E. J. Melero, Radiation-induced void formation and ordering in Ta-W alloys, *J. Nucl. Mater.*, 495 (2017) 343–350.

[114] M. Mathews, B. D. LaFerriere, L. R. Pederson, E. W. Hoppe, Plating of iridium for use as high purity electrodes in the assay of ultrapure copper, *J. Radioanal. Nucl. Chem.*, 307 (2016) 2577–2585.

[115] D. R. Lide, *CRC Handbook of Chemistry and Physics: A Ready-Reference Book of Chemical and Physical Data*, CRC Press, Boca Raton, FL, (1995).

[116] M.-L. Cui, Y.-S. Chen, Q.-F. Xie, D.-P. Yang, M.-Y. Han, Synthesis, properties and applications of noble metal iridium nanomaterials, *Coord. Chem. Rev.*, 387 (2019) 450–462.

[117] R. D. Rovang, M. E. Hunt, Liner protected carbon-carbon heat pipe concept, space nuclear power systems, in *Proceedings of the 9th Symposium*, Part 2, (1992) 781–786.

[118] E. K. Ohriner, Rhenium and iridium, paper presented at Rhenium and Rhenium Alloys, Orlando, FL, 9–13 February (1997)

[119] A. J. Fortini, R. H. Tuffias, The next step in chemical propulsion: Oxide- iridium/rhenium combustion chambers, paper presented at Space Technology and Applications International Forum-1999, the Conference on International Space Station Utilization and Conference on Global Virtual Presence, (1999).

[120] A. J. Fortini, R. H. Tuffias, R. B. Kaplan, A. J. Duffy, B. E. Williams, J. W. Brockmeyer. Iridium/rhenium combustion chambers for advanced liquid rocket propulsion. paper presented at TMS Annual Meeting, TMS, (2000).

[121] C. Stechman, P. Woll, R. Fuller, A. Colette, A high performance liquid rocket engine for satellite main propulsion. paper presented at 36th AIAA/ASME/SAE/ASEE Joint Propulsion Conference, American Institute of Aeronautics and Astronautics, (2000).

[122] I. Zuba, M. Zuba, M. Piotrowski, A. Pawlukojc Ruthenium as an important element in nuclear energy and cancer treatment, *Appl. Radiat. Isot.*, 162 (2020) 109176.

[123] F. E. Poynton, S. A. Bright, S. Blasco, D. C. Williams, J. M. Kelly, T. Gunnlaugsson, The development of ruthenium(ii) polypyridyl complexes and conjugates for in vitro cellular and in vivo applications. *Chem. Soc. Rev.*, 4 (2017) 7706–7756.

[124] C. S. Allardyce, P. J. Dyson, Ruthenium in medicine: Current clinical uses and future prospects, *Platinum Metals Rev.*, 45 (2) (2001) 62–69.

[125] E. A. Seddon, K. R. Seddon, The chemistry of ruthenium, in: R. J. H. Clark (Ed.), *Topics in Inorganic and General Chemistry*, Elsevier, Oxford, (1984) 1–13.

[126] I. C. Smith, B. L. Carson, T. L. Ferguson, Osmium: An appraisal of environmental exposure, *Environ. Health Persp.*, 8 (1974) 201.

[127] P. Walker, W. H. Tarn, *CRC Handbook of Metal Etchants*, CRC Press, Boca Raton, FL, (1990).

[128] G. Girolami, Osmium weighs in, *Nature Chem.*, 4 (2012) 954.

[129] M. C. Green, H. B. Skinner, R. A. Tuck, Osmium-tungsten alloys and their relevance to improved M-type cathodes, *Appl. Surf. Sci.*, 8 (1–2) (1981) 13–35.

[130] A. Ulybkin, A. Rybka, K. Kovtun, V. Kutny, V. Voyevodin, A. Pudov, R. Azhazha, Radiation-induced transformation of hafnium composition, *Nucl. Eng. Technol.*, 51 (8) (2019) 1964–1969.

[131] H. W. Keller, J. M. Shallenberger, D. A. Hollein, A. C. Hott, Development of hafnium and comparison with other pressurized water reactor control rod materials, *Nucl. Technol.*, 59 (3) (1982) 476–482.

[132] O. Levy, G. L.W. Hart, S. Curtarolo, Hafnium binary alloys from experiments and first principles, *Acta. Mater.*, 58 (2010) 2887–2897.
[133] J. A. Davidson, Titanium molybdenum hafnium alloys for medical implants and devices, US Patent 5954724, September 21, (1999).
[134] A. Baudry, P. Boyer, L. P. Ferreira, S. W. Harris, S. Miraglia, L. J. Pontonnier, A study of muon localization and diffusion in Hf_2Co and Hf_2CoH_3., *J. Phys. Condens. Matter*, 4 (1992) 5025.
[135] A. Callegari, et al., Interface engineering for enhanced electron mobilities in W/HfO2 gate stacks, in: *Proceedings of Electron Devices Meeting, IEDM Technical Digest*. IEEE International, Piscataway, NJ, (2004) 825–828.
[136] M. T. Fernandes, Aluminum–magnesium–scandium alloys with hafnium. WO/2001/012868, World Intellectual Property Organization, Geneva, (2001).
[137] M. Modreanu, J. Sancho-Parramon, O. Durand, B. Servet, M. Stchakovsky, C. Eypert, C. Naudin, A. Knowles, F. Bridou, M. F. Ravet, Investigation of thermal annealing effects on microstructural and optical properties of HfO_2 thin films, *Appl. Surf. Sci.*, 253 (2006) 328–334.
[138] J. Vlček, A. Belosludtsev, J. Rezek, J. Houška, J. Čapek, R. Čerstvý, S. Haviar, Highrate reactive high-power impulse magnetron sputtering of hard and optically transparent HfO_2 films, *Surf. Coat. Technol.*, 290 (2016) 58–64.
[139] S. Stojadinović, N. Tadić, R. Vasilić, Plasma electrolytic oxidation of hafnium, *Int. J. Refract. Met. Hard Mater.*, 69 (2017) 153–157.
[140] D. A. Golosov, N. Vilya, S. M. Zavadski, S. N. Melnikov, A. V. Avramchuk, M. M. Grekhov, N. I. Kargin, I. V. Komissarov, Influence of film thickness on the dielectric characteristics of hafnium oxide layers, *Thin Solid Films*, 690 (2019) 137517.
[141] S. Telu, A. Patra, M. Sankaranarayana, R. Mitra, S. K. Pabi, Microstructure and cyclic oxidation behavior of W-Cr alloys prepared by sintering of mechanically alloyed nanocrystalline powders, *Int. J. Refract. Met. Hard Mater.*, 36 (2013) 191–203.
[142] Z. Z. Fang, C. Ren, M. Simmons, P. Sun, The effect of Ni doping on the mechanical behavior of tungsten, *Int. J. Refract. Met. Hard Mater.*, 92 (2020) 105281.
[143] C. Ren, Z. Z. Fang, M. Koopman, B. Butler, J. Paramore, S. Middlemas, Methods for improving ductility of tungsten: A review, *Int. J. Refract Met. Hard Mater.*, 75 (2018) 170–183.
[144] L. B. Lundberg, A critical evaluation of molybdenum and its alloys for use in space reactor core heat pipes, LA-8685-MS, Los Alamos Scientific Laboratory, New Mexico, (1981).
[145] M. S. El-Genk, J-M. Tournier, A review of refractory metal alloys and mechanically alloyed-oxide dispersion strengthened steels for space nuclear power systems, *J. Nucl. Mater.*, 340 (2005) 93–112.
[146] K. Leitner, P. J. Felfer, D. Holec, J. Cairney, W. Knabl, A. Lorich, H. Clemens, S. Primig, On grain boundary segregation in molybdenum materials, *Mater. Des.*, 135 (2017) 204–212.
[147] S. Kalveram, Re-Ni coatings for aircraft and aerospace applications, *Advanced Science News*, April 25, (2016).
[148] S.-Il Baik, A. Duhin, P. J. Phillips, R. F. Klie, E. Gileadi, D. N. Seidman, N. Eliaz, Atomic-scale structural and chemical study of columnar and multilayer Re–Ni electrodeposited thermal barrier coating, *Adv. Eng. Mater.*, 18 (7) (2016) 1133–1144.
[149] D. J. Maykuth, F. C. Holden, R. I. Jaffee, The workability and mechanical properties of tungsten-molybdenum-base alloys containing rhenium, *Electrochemical Society: Symposium on Rhenium*, Elsevier, Oxford, (1962), 114–125.

[150] J. Carlen, B. D. Bryskin, Concerning sigma-phase in molybdenum-rhenium alloys. *J. Mater. Eng. Perform.*, 3 (1994) 282–291.

[151] J. T. Busby, K. J. Leonard, S. J. Zinkle, Radiation-damage in molybdenum–rhenium alloys for space reactor applications, *J. Nucl. Mater.*, 366 (2007) 388–406.

[152] J. A. Richardson, Md. S. H. Bhuiyan, *Corrosion in Hydrogen Halides and Hydrohalic Acids: Reference Module in Materials Science and Materials Engineering*, Elsevier, Oxford, (2017).

[153] Y. Freeman, *Tantalum and Niobium-Based Capacitors Science: Technology, and Applications*, Springer International Publishing, Cham, (2017).

[154] W. P. Wu, Z. F. Chen, Iridium coating: Processes, properties and application. Part I, *Johnson Matthey Technol. Rev.*, 61 (1) (2017) 16–28.

[155] G. Savage, G. M. Savage, E. Savage, *Carbon-carbon Composites*, Chapman & Hall, London, (1993).

[156] H. S. Gandhi, H. K. Stepien, M. Shelef, Optimization of ruthenium- containing, stabilized, nitric oxide reduction catalysts, *Mater. Res. Bull.*, 10 (1975) 837.

[157] J. D. Christian, Process behavior and control of ruthenium and cerium in controlling air-borne effluents from fuel cycle plants, paper presented at The ANS-AIChE Meeting, Sun Valley, Idaho, Aug. (1976).

[158] R. E. Blanco, G. I. Cathers, L. M. Ferris, T. A. Gens, R. W. Horton, E. L. Nicholson, Processing of graphite reactor fuels containing coated particles and ceramics, *Nucl. Sci. Eng.*, 20 (1964) 13.

[159] A. Steffen, K. Bachmann, Gas chromatographic study of volatile oxides and hydroxides of Tc, Re, Ru, Os and Ir, *Talanta*, 25 (1978) 551–556.

[160] F. Garisto, Thermodynamic behaviour of ruthenium at high temperatures, AECL-9552, Atomic Energy of Canada Limited, Pinawa, Manitoba, (1988).

[161] L. Zhou, C. Li, X. Liu, Y. Zhu, Y. Wu, T. von Ree, Metal oxides in supercapacitors, in: Y. Wu (Ed.), *Metal Oxides: Metal Oxides in Energy Technologies*, Elsevier, Oxford, (2018) 169–203.

[162] Anon, Properties of Ruthenium, Osmium and Iridium: Melting point and hardness determinations, *Platinum Metals Rev.*, 4 (1) (1960) 31.

[163] A. S. Darling, The elastic and plastic properties of the platinum metals, *Platinum Metals Rev.*, 10 (1) (1966) 14.

[164] D. R. Holmes, A. W. Chan, Corrosion of hafnium and hafnium alloys, in: S. D. Cramer, B. S. Covino, Jr (Eds.), *ASM Handbook*, 13B, ASM, Materials Park, OH, (2005) 354–358.

[165] D. Gosset, Absorber materials for Generation IV reactors, in: P. Yvon (Ed.), *Structural Materials for Generation IV Nuclear Reactors*, Woodhead Publishing, Cambridge, (2017) 533–567.

[166] C. Suryanarayana, C. C. Koch, Nanostructured materials, in: C. Suryanarayana (Ed.), *Pergamon Materials Series*, 2, Pergamon, Oxford, (1999) 313–344.

[167] H. Gleiter, Nanostructured materials: State of the art and perspectives, *Nanostruct. Mater.*, 6 (1995) 4.

[168] H. Gleiter, Nanocrystalline materials, *Prog. Mater. Sci.*, 33 (1989) 223–315.

[169] M. J. Weins, H. Gleiter, B. Chalmers, Computer calculations of the structure and energy of high-angle grain boundaries, *J. Appl. Phys.*, 42 (1971) 2639.

[170] G. Guisbiers, Advances in thermodynamic modelling of nanoparticles, *Adv Phys: X*, 4 (1) (2019) 1668299.

[171] C. C. Yang, Y.-W. Mai, Thermodynamics at the nanoscale: A new approach to the investigation of unique physicochemical properties of nanomaterials, *Mater. Sci. Eng. R. Rep.*, 79 (2014) 1–40.

[172] W. H. Qi, M. P. Wang, Size and shape dependent melting temperature of metallic nanoparticles, *Mater. Chem. Phys.*, 88 (2004) 280.
[173] J. Shanker, M. Kumar, Studies on melting of alkali halides, *Phys. Stat. Sol. (B)*, 158 (1990) 11.
[174] G. Guisbiers, L. Buchaillot, Modeling the melting enthalpy of nanomaterials. *J Phys. Chem. C.*, 113 (2009) 3566–3568.
[175] F. Calvo, Thermodynamics of nanoalloys, *Phys. Chem. Chem. Phys.*, 17 (2015) 27922–27939.
[176] W. H. Qi, M. P. Wang, Q. H. Liu, Shape factor of nonspherical nanoparticles, *J. Mater. Sci.*, 40 (2005) 2737.
[177] W. H. Qi, B. Y. Huang, M. P. Wang, Z. M. Yin, J. Li, Shape factor for non-cylindrical nanowires, *Phys B: Condens. Matter.*, 403 (2008) 2386.
[178] J. A. Vanvechten, M. Wautelet, Variation of semiconductor band-gaps with lattice temperature and with carrier temperature when these are not equal. *Phys Rev. B*, 23 (1981) 5543–5550.
[179] M. Li, J. C. Li, Size effects on the band-gap of semiconductor compounds, *Mater. Lett.*, 60 (2006) 2526–2529.
[180] M. Goyal, M. Singh, Size and shape dependence of optical properties of nanostructures, *Appl. Phys. A*, 126 (176) (2020).
[181] T. Vossmeyer, L. Katsikas, M. Giersig, I. G. Popovic, K. Diesner, A. Chemseddine, A. Eychmueller, H. Weller, CdS nanoclusters: Synthesis, characterization, size dependent oscillator strength, temperature shift of the excitonic transition energy, and reversible absorbance shift, *J. Phys. Chem.*, 98 (1994) 7665–7673.
[182] R. Banerjee, R. Jayakrishnan, P. Ayyub, Effect of the size-induced structural transformation on the band gap in CdS nanoparticles, *J. Phys-Condens. Matter.*, 12 (2000) 10647–10654.
[183] T. Torimoto, H. Kontani, Y. Shibutani, S. Kuwabata, T. Sakata, H. Mori, H. Yoneyama, Characterization of ultrasmall CdS nanoparticles prepared by the size-selective photoetching technique, *J. Phys. Chem. B.*, 105 (29) (2001) 6838–6845.
[184] N. Soltani, E. Gharibshahi, E. Saion. Band gap of cubic and hexagonal CdS quantum dots – experimental and theoretical studies, *Chalcogenide Lett.*, 9 (2012) 321–328.
[185] G. Guisbiers, M. Kazan, O. V. Overschelde, M. Wautelet, S. Pereira, Mechanical and thermal properties of metallic and semiconductive nanostructures, *J. Phys. Chem. C*, 112 (11) (2008) 4097–4103.
[186] I. A. Ovidko, R. Z. Valiev, Y. T. Zhu, Review on superior strength and enhanced ductility of metallic nanomaterials, *Prog. Mater. Sci.*, 94 (2018) 462–540.
[187] M. F. Ashby, The deformation of plastically non-homogeneous materials, *Philos. Mag.*, 21 (170) (1970) 399–424.
[188] H. J. Gao, Y. G. Huang, W. D. Nix, J. W. Hutchinson, Mechanism-based strain gradient plasticity–I. Theory, *J. Mech. Phys. Solids*, 47 (1999) 1239–1263.
[189] G. Liu, G. J. Zhang, F. Jiang, X. D. Ding, Y. J. Sun, J. Sun, E. Ma, Nanostructured high-strength molybdenum alloys with unprecedented tensile ductility, *Nature Mater.*, 12 (4) (2013) 344–350.
[190] V. S. Saji, R. Cook, *Corrosion Protection and Control Using Nanomaterials*, Woodhead Publishing Limited, Cambridge, (2012).

2 Principles of Oxide Dispersion Strengthening

2.1 CHARACTERISTICS OF OXIDES

Oxides have exciting properties, such as ferroelectric, dielectric piezoelectric, fuel cell, magnetic, etc., and varied applications. They can be classified into several categories based on metal or nonmetal, mixed oxides (Perovskite oxides: two different cations), crystal and electronic structure, density, etc. Depending on the partial pressure of oxygen and temperature, many oxides are non-stoichiometric. The oxides formula M_xO_y can be changed to $M_{x(1-a)}O_y$ when the x, y values are different and the value can be varied significantly by a different percentage. The presence of metal vacancies can agglomerate, and a compound defect structure can be formed, resulting in a reduction of the free energy of the system [1, 2]. In several systems, such as TiO_2, WO_3, Nb_2O_5, the non-stoichiometry is adjusted through planar defects [1]. The chapter also indicates that stoichiometric deviation is arranged in ascending order between several metal oxides as: FeO>MnO>CoO> NiO [1]. Table 2.1 shows different oxides and their characteristics [3].

Oxides with a semi-covalent nature exhibit elevated O $1s$ binding energy ($\approx$ 533.5–530.5 eV) and this is related to considerable and different covalency. Ionic oxides with a medium-range O 1s binding energy ($\approx$ 530±0.4 eV) exhibit 76–89% ionicity, whereas highly ionic oxides with low O $1s$ binding energy ($\approx$ 529.5±528.0 eV) show > 90% ionicity [3]. The optical basicity is inversely proportional to the interaction parameter. A reduced interaction parameter and an increase in optical basicity lead to enhancement of the electron density [3]. Figure 2.1 shows the rare earth cubic C-type yttrium oxide (Y_2O_3) (bixbyite) structure, the sites of the oxygen atoms (48e), and the Y atoms (Y_1: 8b, Y_2: 24d) are presented [4]. According to the literature, the diffusion-related energy barrier for rare earth oxides is the smallest [5]. Therefore, intense diffusion results in a reduction of oxide free energy and novel phases are stabilized [4].

Figure 2.2 presents the FCC structure in the X_2O composition (X = Li, Na, K, Rb), space group Fm$\bar{3}$m and various symmetries of cation and anion sublattice [6, 7]. In the structure the oxygen ions are positioned related to the FCC structure, whereas the metal ions are positioned to generate simple cubic structures [7]. Benco has illustrated the covalent bonding in MgO and Y_2O_3 [8]. The covalency degree of Y_2O_3 is comparatively higher than MgO [8]. The melting point of the oxides depends on the band gap and a reduction in the band gap in Y_2O_3 results in a reduced melting temperature compared to MgO. The high temperature applicability of oxides depends on their stability and Gibbs free energy (standard) at elevated temperatures. A comparative study of Al_2O_3, TiO_2, Nb_2O_5, VO_2, Cr_2O_3 shows that the Gibbs free energy (standard) becomes less negative (reduced stability) with an increase in temperature

TABLE 2.1
Classification of Simple Oxides

Oxides	$\alpha_{O^{2-}}$ (Å³)	E_b (ev)	α_1 (Å³)	E_B (ev)	Λ	Λ (Å⁻³)	Bonding
Semivalent (predominantly acidic oxides) BeO, B_2O_3, P_2O_5, SiO_2, Al_2O_3, MgO, GeO_2, Ga_2O_3	≈ 1–2 Low	≈ 533.5–530.5 High	≈ 0.032–0.2 Low	≈ 60–200 High	≈ 0.3–0.7 ≈ 0.26–0.11 Acidic Strong interionic interaction		Large overlap between O 2p and valence metal orbitals	Strong covalent bonds
Ionic (basic) Li_2O, CaO, Sc_2O_3, TiO_2, V_2O_5, MnO, Fe_2O_3, CoO, NiO, CuO, ZnO, Y_2O_3, ZrO_2, Nb_2O_5, MoO_3, In_2O_3, SnO_2, TeO_2, CeO_2, Ta_2O_5, WO_3	≈ 2–3 High	≈ 530±0.4 Medium range	≈ 0.2–0.8 High	≈ 15–60 Low	≈ 0.8–1.1 ≈ 0.11–0.03 Basic mainly weak interionic interaction		Smaller overlap between O 2p and metal valence orbitals	Bonds with increased ionicity
Very ionic (very basic) Na_2O, SrO, CdO, Sb_2O_3, Cs_2O, BaO, PbO, Bi_2O_3	≈ >3 Very high	≈ 529.5–528.0 Low	≈ 0.8–3.7 Very high	≈ <20 Very low	≈ >1.1 ≈ <0.03 Very basic Very weak interionic interaction		Small overlap between O 2p and metal valence orbitals	Very ionic bonds

Source: Reprinted from [3]. Copyright (2002), with permission from Elsevier.

Note: Group of oxides, oxide ion polarizability (αO^{2-}), O 1s binding energy (Eb), cation polarizability ($\alpha 1$), metal (or nonmetal) binding energy (EB), optical basicity (Λ), interaction parameter (Λ), and type of bonding

Principles of Oxide Dispersion Strengthening

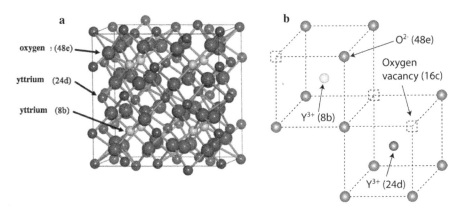

FIGURE 2.1 Cubic C-type Y_2O_3 (bixbyite) structure. The lattice parameter is $a_0 = 1.0604$ nm. (a) Unit cell representation. (b) Environment of the two Y sites (the actual position of the O atoms is slightly shifted from the corners of the cubes). Structural oxygen vacancies are denoted by small squares. (Reprinted from [4]. Copyright (2016), with permission from Elsevier.)

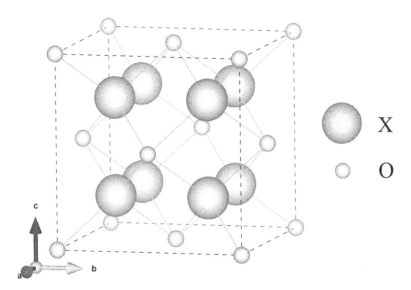

FIGURE 2.2 FCC crystal structure of X_2O (X = Li, Na, K and Rb). (Reprinted from [6]. Copyright (2020), with permission from Elsevier.)

from 800°C to 1200°C [9]. The literature has illustrated that maximum stability is attained in Al_2O_3 irrespective of temperature [9].

The mechanical properties of oxides also influence the properties of the refractory metals in oxide dispersion strengthened alloys. The strength of the refractory metals with added oxides should not deteriorate with an increase in temperature. Some mechanical properties of the oxides and the impact of temperature on mechanical properties are presented in Table 2.2.

TABLE 2.2
Mechanical Properties of Oxides

Oxide	Temperature (°C)	Tensile strength MN/m²	Compressive strength (MN/m²)	Bending strength MN/m²
Al_2O_3	20	258.622 [10]	2929.22 [10]	156.80
	800	234.612 [10]	240.10 [10]	(Calcined at 1750°C) [17]
	1200	126.91 [10]		
	1400	29.498 [10] (sintered)		
TiO_2	20	58.80–68.60 [11] Density: 4100 Kg/m³	245 [11] Density: 4100 Kg/m³	Temperature: - 134.26 (Density: 4100 kg/m³) [11]
ZrO_2	20	145.53 [12, 13]	2058 [12, 14]	6.89 (at 1027°C)
	1000	93.10 [12, 13]	1176 [12, 14]	(under condition of apparent porosity of 18.6%) [18]
	1200	82.52 [12, 13] stabilized with 4.2% (by mass) MgO 104,105	784 [12, 14] (sintered) 104,282	
ThO_2	20	82.32 [12, 15, 16]	1372 [12,14]	82.32 [19]
	400	68.60 [12, 15, 16]	1078 [12,14]	123.48 [20]
	1000	48.02 [12, 15, 16] Porosity: 3–7%	352.8 [12, 14] Sintered at 2173°K	

Sources: [10–20].

Principles of Oxide Dispersion Strengthening

It is evident from Table 2.2 that the mechanical properties of oxide materials depend on the application temperature, the processing conditions, such as the porosity percentage, the stabilization effect, sintering and the calcination effect. Therefore, the impact of oxides on the properties of refractory metals depends on the application environment and thermal stability. Oxides should possess high thermal stability (no dissociation or oxygen release) at elevated temperature applications. Oxides such as Al_2O_3, Cr_2O_3, ZrO_2, Y_2O_3, ThO_2, La_2O_3, Eu_2O_3, CeO_2, Nd_2O_3, HfO_2 are thermally stable, whereas HgO, AuO, IrO_2, CdO_2, RuO_2, ZnO, CuO are unstable oxides [21].

2.2 CRITERIA FOR THE SELECTION OF OXIDES

As discussed in Section 2.1, the properties of oxides, such as melting point, high temperature strength, high temperature stability, density of oxides, thermal expansion coefficient and molar volume at the application temperature, should be the basis for selection when adding in refractory metals. High temperature strength and stability are of prime importance to restrain grain growth and improve the matrix's strength. Rare earth metals, such as Y, La, have a strong interaction with oxygen and produce stable oxides compared to Cr_2O_3, Al_2O_3, TiO_2, ZrO_2 [22]. Rare earth metals combine with oxygen and purify the grain boundary and improve the grain boundary bonding with the matrix. Oxides should not be volatile (no sublimation), any undue phase transformation which significantly changes the volume needs to be restricted. The vapor pressure of the added oxide is another important factor, e.g., the vapor pressure of Y_2O_3 at 2327°C is 0.162 (oxide) N/m^2 and for ZrO_2 at the identical temperature it is 0.08 N/m^2 (oxide) [23, 24]. The Pilling-Bedworth ratio (volume of the oxide:volume of metal) is also essential in the presence of dispersed oxides in refractory metals for high temperature applications. If the ratio is <1, such as in the case of alkali metals and alkaline earth metals, this causes inadequate inhibition of oxidation and tensile stress develops, for the ratio > 2, compressive stress develops. However, for some refractory oxides, such as Ta_2O_5, the ratio is 2.44 but prevents pitting corrosion [25]. During elevated temperature usage, the metals and oxides are subjected to expansion. A significant difference in thermal expansion coefficient can lead to cracks or void initiation. However, the literature shows that the rare earth oxide addition (Y_2O_3) in an W-Cr alloy system can resist crack propagation, improve the adhesion of the oxide scale, and enhance the oxidation resistance [26].

2.3 DISPERSION STRENGTHENING

In the presence of precipitates, dislocation spreads through the precipitates and therefore increases the strength of the alloy, provided the precipitates are incoherent (lattice mismatch with the matrix phase) and hard. so that it is preferable to by-pass rather than cut off precipitates by dislocation. This phenomenon leads to increased dislocation density and an increased rate of strain hardening. The lattice misfit between the precipitate and the matrix inhibits the movement of the dislocation and results in strength enhancement. If the particles are coarser, the strength of the alloy is reduced. Therefore, the dispersion needs to be uniform in the matrix, and a critical size and

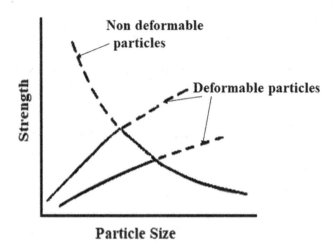

FIGURE 2.3 Change in strength with particle (deformable and non-deformable) size. (Reprinted from [27]. Copyright (2014), with permission from Elsevier.)

concentration are required for effective strengthening [27]. Figure 2.3 shows the change in strength with the particle size.

The curvature radius with respect to dislocation bending (r) is related to shear modulus (µ), Burgers vector (b), applied stress (τ) as:

$$r = \frac{\alpha \mu b}{\tau} \tag{2.1}$$

where α is a factor = ½.

During the aging process, when the precipitate size increases, the strength decreases due to the enhancement of the obstacle distance. Nano precipitates that are insoluble in the matrix can enhance the resistance against creep. For such a condition the strengthening can be illustrated by the Orowan mechanism. The Orowan strengthening ($\Delta\sigma_{orowan}$), as expressed by the Orowan–Ashby equation, is [28]:

$$\Delta\sigma_{orowan} = \frac{0.13 G_m b}{\lambda} \ln\frac{r}{b} \tag{2.2}$$

where G_m is the shear modulus of the matrix, b denotes the matrix's Burgers vector, λ represents the interparticle spacing, r denotes the particle radius = $d_p/2$ (d_p is particle diameter). The interparticle spacing λ is related to the particle diameter and volume fraction of the nanoparticles (V_p) used for reinforcing as [29, 30]:

$$\lambda \approx d_p \left[\left(\frac{1}{2V_p}\right)^{1/3} - 1\right] \tag{2.3}$$

Principles of Oxide Dispersion Strengthening

The literature also reports that the dislocation density increment owing to the variance ($\Delta \alpha$) in the thermal expansion coefficient among particles and the matrix after fabrication (a cooling regime) is related to V_p, d_p, b, $\Delta\alpha$ and ΔT as [31]:

$$\rho = 12 \frac{\Delta\alpha \Delta T V_p}{bd_p (1-V_p)} \quad (2.4)$$

where ΔT is the variance between the fabrication and the test temperature.

The d spacing of the matrix is influenced by the presence of solute elements or precipitate oxide particles which leads to lattice misfit. The lattice misfit provides significant insight into the coherency strain between the matrix and the precipitates or solute elements. Reports suggest the variation of the lattice parameter of the isotropic matrix (Δa_a) (without strain) in the presence of isotropic inclusions depends on the misfit parameter (ε) as [32–34]:

$$\Delta a_a = \frac{4\mu_A C_6}{3K_A} \frac{\varepsilon}{(1+\varepsilon)^3} y_B a_A \quad (2.5)$$

$$C_6 = \frac{3K_B}{3K_B + 4\mu_A} \quad (2.6)$$

where a_a is the lattice parameter of A, K is the bulk modulus of inclusions, μ is the shear modulus of inclusions, y_B is the volume fraction of inclusions.

The extent of strengthening due to oxide dispersion depends on the volume fraction (as discussed above), the size and the shape of the oxide particles. Finer size and higher volume fraction enhance the pinning pressure on the grain boundary and contribute to strengthening of the alloy. However, uniform dispersion of the oxide particle is of paramount importance to achieve the isotropic properties of the material.

2.4 MECHANISM OF OXIDE DISPERSION STRENGTHENING

The mechanism of dislocation movement around dispersed oxides varies with the change in particle size and the interparticle spacing. The dislocation loop formation during by-passing of dislocation across the precipitate particles is schematically illustrated in Figure 2.4. This type of dislocation bending occurs during large spacing between precipitates. The plastic flow in the case of non-deformable particles is turbulent in nature and it leads to an increase in dislocation loops and dipoles density [27].

The strengthening during oxide dispersion is associated with the Zener Pinning effect. During the interaction of a spherical particle with the grain boundary an attractive force (the pinning force) is operative, which acts against the boundary dragging on the force. During the boundary-particle interaction, the grain boundary area decreases and furthermore, the grain boundary area increases to remove the particle, which is made possible by increasing the energy of the system. The force related to the pinning, considering the particle incoherency is [35]:

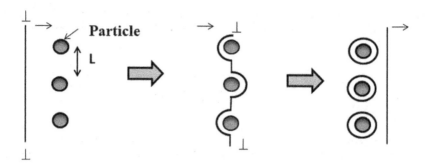

FIGURE 2.4 Dislocation bowing mechanism among precipitates with larger spacing [27].

$$F = 2\pi r \gamma \cos\theta \sin\theta \qquad (2.7)$$

where γ is the grain boundary surface tension, θ is the bypass angle. When $\theta = \pi/4$, the maximum force of pinning is given by:

$$F_p = \pi r \gamma \qquad (2.8)$$

The number of particles per unit volume N is presented as:

$$N = \frac{3F}{4\pi r^3} \qquad (2.9)$$

In case of planar boundary, the particle fraction related to the boundary contact is:

$$N_i = 2RN = \frac{3F}{2\pi r^2} \qquad (2.10)$$

The pinning pressure is:

$$(P_p) = F_p \times N_i \qquad (2.11)$$

or,

$$P_p = \frac{3\gamma F}{2r} \qquad (2.12)$$

Figure 2.5 shows the particle interaction with the boundary [36]. The pinning particles' location changes owing to the moving boundary and, therefore, the pinning pressure exerted by the particles also changes. During the particle growth an incoherent boundary with elevated energy develops along with relieving of the elastic strain energy. Elastic energy and interfacial energy are present in the incoherent interface. Both types of energy are enhanced with an increase in the particle size. However, the dominant effect is the elastic energy with respect to particle growth. The lattice misfit is significant in coarse particles. The pinning pressure reduces for

Principles of Oxide Dispersion Strengthening

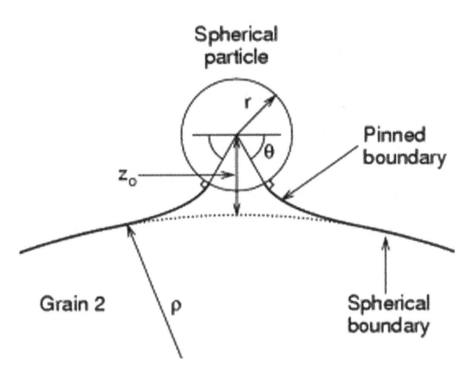

FIGURE 2.5 A schematic of a grain boundary by-passing a particle: r is the particle radius, ρ is the radius of boundary curvature, θ is the boundary bypass angle and z_0 is the distance of the boundary from the particle center. The boundary migrates with a right angle at the boundary/particle interface. (Reprinted from [36]. Copyright (2006), with permission from Elsevier.)

larger particles and higher misfit owing to the enhanced elastic energy compared to the matrix-particle interfacial energy [36–38]. Reports indicate that the lattice misfit is 5% when the grain boundary energy is almost 1 J/m² and the elastic modulus is 100 GPa with the grain size of 10 nm [38]. For the same grain boundary energy and elastic modulus, the lattice misfit increases to 10% with an increase in particle size to 30 nm and the magnitudes of elastic and interfacial energy are 10^{-6} and 10^{-7} respectively [38].

2.5 SUITABILITY OF NANO-OXIDES OVER MICRON-SIZED OXIDE DISPERSION

As discussed in Section 2.4, the oxide dispersion size is significant in controlling the grain growth during heat treatment. During heat treatment, it is anticipated that some growth of the dispersed particles will occur. During powder processing of ODS alloys, micron-sized powder shows less tendency to agglomeration compared to nano-sized powder, due to less surface energy. The strength of fracturing also reduces with an increase in particle size. However, for an effective size reduction in the matrix powder, it is important to increase the powder processing energy to reduce the consolidation temperature. Nano-sized powder contributes considerable energy during the reduction in the matrix particle size. The strength of the ODS refractory alloys is contributed by grain size strengthening and particle strengthening. Reports indicate that the effect of alloy strengthening is comparatively higher due to the particle contribution rather than grain size strengthening [39]. It is also reported that the yield strength and tensile strength of Mo-La_2O_3 alloy show less enhancement when the particle size increases beyond 82.3 nm, as well as when the Orowan stress increases with the increase in the volume fraction of the particles [39]. However, one of the significant challenges with finer microstructures is controlling the ODS alloys' shrinkage during prolonged heat treatment. Apart from the shrinkage, the high cost of nano-oxides adds to the overall processing cost increment. Cai et al. have reported that the grain refinement of sintered WC-Co alloys is higher with lower wt.% (0.4%) nano-oxide (CeO_2) against micron-sized dispersion [40]. However, the average grain size increases from 410 nm to 470 nm with an increase in wt.% of CeO_2 from 0.4% to 1.2%, owing to interfacial segregation; still, nano-oxide dispersion in low concentration (0.4 wt.%) shows an improvement in densification, hardness, fracture toughness, and wear resistance [40]. The literature shows that the crystallite size of Fe, during mechanical alloying Fe–12Cr–2W–0.5Y_2O_3 alloy with micron-sized (45 μm) and nano-sized (25 nm) Y_2O_3 dispersion, exhibits enhanced reduction with micron-sized Y_2O_3 dispersion owing to less pinning action on the dislocation motion [41]. However, the consolidated grain size is less and the creep resistance is higher in nano Y_2O_3 dispersed alloys than micro Y_2O_3 dispersed alloys, due to the effective inhibition of the grain growth and the dislocation movement [41]. Daoush et al. have reported that finer oxide dispersion reduces the grain size in sintered alloys and improves the hardness and compressive strength of W-Ni-TiO_2 alloy, compared to W-Ni-Y_2O_3, W-Ni-ZrO_2 alloy [42]. It is also illustrated that finer grain size results in increased uniformity in grain size distribution, enhanced densification and homogeneous structure of indium tin oxide ceramic [43]. An increase in the size of the oxide particles at the grain boundary also incorporates stress concentration followed by strain localization which leads to cracking and reduces ductility [44].

2.6 EFFECTIVENESS OF DISPERSION STRENGTHENING BY OXIDES

Significantly, high temperature refractory metals should retain their strength at elevated temperatures without any softening. However, recrystallization behavior significantly limits the high temperature properties, such as strength and creep resistance. Dispersion strengthening has a considerable impact on restricting the recrystallization of the refractory metals by impeding the grain boundary sliding [45–48]. Recrystallization-based embrittlement of W and Mo alloys reduces the resistance against irradiation in nuclear applications and decreases the elongation consistency and enhancement in DBTT [49]. Cost-effective machining of W is possible at room temperature when La_2O_3 is added. Improvement in resistance to creep, thermal shock resistance, hot tensile strength, and no occurrence of cracking during deformation of the alloy are reported [50]. These properties lead to the application of $W-La_2O_3$ alloys in the baffle of the International Thermonuclear Experimental Reactor (ITER) [51]. Dispersed particles are selected based on the inertness of the size due to the heat treatment. Dissolution of solute precipitate particles during high temperature applications leads to a reduction in the strength [52]. The alloys with dispersion strengthening can retain strength until 90% of the melting point of the alloy system [52]. A ThO_2 dispersed W alloy possesses enhanced strength until ~ 2315.56°C (4200°F) compared to solid solution strengthening or precipitation strengthening [53]. Martinez et al. have fabricated $W-Ti-La_2O_3$, $W-V-La_2O_3$, $W-V-Y_2O_3$ and reported that the alloys have appreciable density and hardness and possess the potential to restrain crack growth [54]. Extended fatigue life and improvement in fatigue ductility are also reported in ODS-Mo alloys [55]. ODS-Mo alloys exhibit enhanced creep rupture strength compared to arc cast Mo, arc cast W, powder metallurgically processed Mo-50Re alloy and Re at 1600°C [56–58]. The DBTT of swaged ODS-Mo (-60°C) is reduced compared to powder metallurgically processed swaged Mo (-23°C) [56]. The fracture toughness of ODS Mo is 52.8 MPa√m compared to commercially pure Mo (24.2 ±2.3 MPa√m) [59, 60]. Modification of the grain boundary or the phase boundary can be achieved by oxide dispersion. The presence of a higher fraction of the grain boundary region with an impurity concentration facilitates the migration of the impurities in the matrix phase and enhances the strength [61]. Increased grain boundary also provides the sink for the defects generated by irradiation, enhancing the resistance against irradiation [62]. However, the effectiveness of oxide dispersion is appreciable if some of the oxide particles are driven from the boundary to the grain matrix to enrich work hardening and uniform elongation [61]. One of the key requirements for a kinetic penetrator alloy is the self-sharpening effect which increases the penetration depth. Further, during ballistic application under high strain rate, thermal softening occurs, which hastens the localized deformation and results in adiabatic shear band formation [63]. It is evident from Figure 2.6 that the presence of precipitate particles (high strength μ phase) induces self-sharpening behavior in equimolar W-Ni-Fe-Mo high entropy alloy compared to 93W-4.9Ni-2.1Fe alloy [64]. The depth of penetration of W-Ni-Fe-Mo high entropy alloy is comparatively higher than 93W-4.9Ni-2.1Fe alloy (Figure 2.6 (a)) [64]. The paper also illustrates that variation in deformation between the matrix and the precipitate particles triggers dynamic recrystallization softening, which contributes to shear band formation and the self-sharpening effect [64]. Finer grain 95W-3.75Ni-1.25Fe alloy exhibits enhanced penetration depth and self-sharpening compared to conventional 95W-3.75Ni-1.25Fe alloy, owing to reduced failure strain in the elevated strain rate condition [65]. The literature states that a mushroom-head formation in liquid phase sintered W-Ni-Fe alloys reduces the penetration efficiency, whereas self-sharpening is evident in Y_2O_3 dispersed W-Ni-Fe alloys [66]. The self-sharpening is attributed to de-bonding of the W/W interface rather than shear band formation in the ODS-W alloy which may improve the applicability of the penetrator [66].

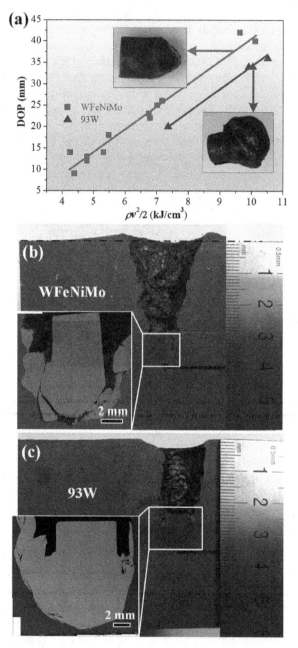

FIGURE 2.6 (a) Depth of penetration of WFeNiMo rod and 93 W rod versus kinetic energy per volume calculated by $\rho v^2/2$, with photographs of the retrieved remnants, respectively; longitudinal sections of medium carbon steel targets impacted by (b) a WFeNiMo penetrator and (c) a 93 W penetrator, with SEM micrographs of the remnant in the corresponding insets, respectively. (Reprinted from [64]. Copyright (2020), with permission from Elsevier.)

2.7 INFERENCES

Refractory metals are potential candidates for high temperature structural applications, as discussed in Chapter 1. However, several limitations, such as poor oxidation resistance of some refractory metals, such as W, Mo, inadequate strength, and creep resistance recrystallization at elevated temperature, significantly reduce their effectiveness. Dispersing oxides in refractory metals is a significant approach to increase the recrystallization temperature, the creep rupture life, the elevated temperature strength and resistance to high temperature oxidation, which are vital in nuclear reactor and defense systems applications. The dispersed oxides restrict the grain growth at high temperatures and facilitate the enhancement of strength and creep resistance. The location of the dispersed oxides is also significant to achieve a synergy between strength-ductility. Higher volume fraction, coarse-grained oxide particles at the boundary increase the interfacial de-cohesion and crack propagation through the grain boundary. Dispersing a fraction of the oxide particles inside the grain, rather than the presence at the boundary is an important strategy to develop bimodal grain size distribution which improves the toughness of the alloy.

REFERENCES

[1] C. Monty, Diffusion in stoichiometric and non-stoichiometric cubic oxides, *Radiation Effects*, 74 (1–4) (1983) 29–55.
[2] O. Paladino, Defect structure of metal oxides at high temperatures, *Chem. Eng. Technol.*, 16 (1993) 341–346.
[3] V. Dimitrov, T. Komatsu, Classification of simple oxides: A polarizability approach, *J. Solid State Chem.*, 163 (1) (2002) 100–112.
[4] R. J. Gaboriaud, F. Paumier, B. Lacroix, Disorder–order phase transformation in a fluorite-related oxide thin film: In-situ X-ray diffraction and modelling of the residual stress effects, *Thin Solid Films*, 601 (2016) 84–88.
[5] L. Kittiratanawasin, R. Smith, B. P. Uberuaga, K. E. Sickafus, Radiation damage and evolution of radiation-induced defects in Er_2O_3 bixbyite, *J. Phys. Condens. Matter*, 21 (2009) 115403.
[6] X. Wu, Y. Zhang, J. Zhang, R. Liu, J. Yang, B. Yang, H. Xu, Y. Ma, High pressure X-ray diffraction study of sodium oxide (Na_2O): Observations of amorphization and equation of state measurements to 15.9 GPa, *J. Alloys Compd.*, 823 (2020) 153793.
[7] F. A. Shunk, *Constitution of Binary Alloys* (2nd Suppl.), McGraw-Hill, New York, (1969).
[8] L. Benco, Covalent bonding in MgO and Y_2O_3, *Ceram. Int.*, 21 (1995) 283–287.
[9] J. Zheng, X. Hou, X. Wang, Y. Meng, X. Zheng, L. Zheng, Isothermal oxidation mechanism of a newly developed Nb-Ti-V-Cr-Al-W-Mo-Hf alloy at 800–1200°C, *Int. J. Refract. Met. Hard Mater.*, 54 (2016) 322–329.
[10] E. Ryshkewitch, BeR, *Deutsch. Keram. Ges.*, 22 (363) (1941) 54.
[11] M. M. Khrushchov, E. S. Berkovich, Microhardness, in *Proc. of Conference on Microhardness*, Izd. Akad. Nauk SSSR, Moscow, (1951) [in Russian].
[12] Anon. *Handbook of Refractory Production*, Izd. Metallurgiya, Moscow, (1965) [in Russian].
[13] P. P. Budnikov (Ed.), *The Physicochemical Bases of Ceramics.: A Collection of Articles*, Gosstroiizdat, Moscow, (1956) [in Russian].

[14] E. Ryshkewitch, *Oxydkeramik der Einstoffsysteme vom Standpunkt der physikalischen Chemie*, Springer-Verlag, Berlin (1948).
[15] W. D. Kingery, Factors affecting thermal stress resistance of ceramic materials, *J. Amer. Ceram. Soc.*, 38 (3) (1955) 3–15.
[16] W. D. Kingery, *Property Measurements at High Temperatures*, Wiley, New York, (1959).
[17] A. I. Avgustinik, *Physicochemical Bases of Ceramics*, Goskhimizdat, Moscow, (1956) [in Russian].
[18] I. S. Mayauskas, A. Ya. Peras, *Zavod. Lab.*, 32 (1966) 885.
[19] J. R. Johnson, S. D. Fulkerson, A. J. Taylor, Technology of uranium dioxide: A reactor material, *Amer. Ceram. Soc. Bull.*, 36 (1957) 112.
[20] M. D. Burdick, H. S. Parker, Effect of particle size on bulk density and strength properties of uranium dioxide specimens, *J. Amer. Ceram. Soc.*, 39 (1956) 181.
[21] G. V. Samsonov, *The Oxide Handbook*, translated from Russian by C. N. Turton and T. I. Turton, Plenum Press, London (1973).
[22] T. H. Okabe, C. Zheng, Yk. Taninouchi, Thermodynamic considerations of direct Oxygen removal from Titanium by utilizing the deoxidation capability of rare earth metals, *Metall. Mater. Trans. B*, 49 (2018) 1056–1066.
[23] E. McCafferty, *Introduction to Corrosion Science*, Springer, Berlin (2010).
[24] M. M. Nakata, R. L. McKisson, B. D. Pollock, Vaporization of Zirconium oxide, *Atomics International*, (1961) 1–10.
[25] Anon, *Thermodynamic Properties of Individual Substances*, vols. 1 and 2, Izd. Akad. Nauk SSSR, Moscow, (1962) [in Russian].
[26] S. Telu, R. Mitra, S. K. Pabi, Effect of Y_2O_3 Addition on oxidation behavior of W-Cr alloys. *Metall. Mater. Trans. A*, 46 (2015) 5909–5919.
[27] R. E. Smallman, A. H. W. Ngan, Precipitation hardening, in: R. E. Smallman, A. H. W. Ngan (Eds.), *Modern Physical Metallurgy* (8th edn), Butterworth-Heinemann, Oxford, (2014) 499–527.
[28] G. E. Dieter, *Mechanical Metallurgy* (3rd edn), McGraw-Hill, New York, (1986) 212–220.
[29] M. A. Meyers, K. K. Chawla, *Mechanical Behaviour of Materials*, Prentice Hall, Upper Saddle River ,NJ, (1999) 492–494.
[30] Q. Zhang, D. L. Chen, A model for predicting the particle size dependence of the low cycle fatigue life in discontinuously reinforced MMC, *Scripta Mater.*, 51 (2004) 863–867.
[31] M. Taya, R. J. Arsenault, *Metal Matrix Composites: Thermomechanical Behaviour*, Pergamon Press, New York, (1989).
[32] J. W. Christian, *The Theory of Transformations in Metals and Alloys*, Pergamon Press, Oxford, (2002).
[33] E. J. Mittemeijer, P. van Mourik, T. H. de Keijser, Unusual lattice parameters in two-phase systems after annealing, *Philos. Mag. A*, 43 (1981) 1157–1164.
[34] M. Akhlaghi, T. Steiner, S. R. Meka, E. J. Mittemeijer, Misfit-induced changes of lattice parameters in two-phase systems: Coherent/incoherent precipitates in a matrix, *J. Appl. Crystallogr.*, 49 (1) (2016) 69–77.
[35] C. S. Smith, Introduction to grains, phases, and interfaces: An interpretation of microstructure, *Trans. AIME*, 175 (1948) 15.
[36] A. Harun, E. A. Holm, M. P. Clode, M. A. Miodownik, On computer simulation methods to model Zener pinning, *Acta Mater.*, 54 (12) (2006) 3261–3273.
[37] T. Chakrabarti, S. Manna, Zener pinning through coherent precipitate: A phase-field study, *Comput. Mater. Sci.*, 154 (2018) 84–90.

[38] N. Wang, Y. Ji, Y. Wang, Y. Wen, L. Q. Chen, Two modes of grain boundary pinning by coherent precipitates, *Acta Mater.*, 135 (2017) 226–232.
[39] G.-J. Zhang, Y.-J. Sun, R.-M. Niu, J. Sun, J.-F. Wei, B.-H. Zhao, L.-X. Yang, Microstructure and strengthening mechanism of oxide lanthanum dispersion strengthened molybdenum alloy, *Adv. Eng. Mater.*, 6 (2004) 943–948.
[40] H. Cai, W. Jing, S. Guo, L. Liu, Y. Ye, Y. Wen, Y. Wu, S. Wang, X. Huang, J. Zhang, Effects of micro/nano CeO_2 on the microstructure and properties of WC-10Co cemented carbides, *Int. J. Refract. Met. Hard Mater.*, 95 (2021), 105432.
[41] P. Susila, D. Sturm, M. Heilmaier, B. S. Murty, V. S. Sarma, Effect of yttria particle size on the microstructure and compression creep properties of nanostructured oxide dispersion strengthened ferritic (Fe-12Cr-2W-0.5Y_2O_3) alloy, *Mater. Sci. Eng. A*, 528 (13–14) (2011) 4579–4584.
[42] W. M. R. Daoush, A. H. A. Elsayed, O. A. G. E. Kady, M. A. Sayed, O. M. Dawood, Enhancement of physical and mechanical properties of oxide dispersion-strengthened tungsten heavy alloys, *Metall. Mater. Trans. A*, 47 (2016) 2387–2395.
[43] F. Mei, T. Yuan, R. Li, L. Zhou, K. Qin, Effects of particle size and dispersion methods of In_2O_3-SnO_2 mixed powders on the sintering properties of indium tin oxide ceramics, *Int. J. Appl. Ceram. Technol.*, 15 (2018) 89–100.
[44] P. M. Cheng, G. J. Zhang, J. Y. Zhang, G. Liu, J. Sun, Coupling effect of intergranular and intragranular particles on ductile fracture of Mo-La_2O_3 alloy, *Mater. Sci. Eng. A*, 640 (2015) 320–329.
[45] P. K. Wright, The high temperature creep behavior of doped tungsten wire, *Metall. Trans. A*, 9 (7) (1978) 955–963.
[46] J. W. Pugh, On the short time creep rupture, *Metall. Trans.*, 4 (2) (1973) 533–538.
[47] H. J. Ryu, S. H. Hong, Fabrication and properties of mechanically alloyed oxide-dispersed tungsten heavy alloys, *Mater. Sci. Eng. A*, 363 (1–2) (2003) 179–184.
[48] M. Mabuchi, K. Okamoto, N. Saito, T. Asahina, T. Igarashi, Deformation behavior and strengthening mechanisms at intermediate temperatures in W-La_2O_3, Mater. *Sci. Eng. A*, 237 (1997) 241–249.
[49] M. S. El-Genk, J.-M. Tournier, A review of refractory metal alloys and mechanically alloyed-oxide dispersion strengthened steels for space nuclear power systems, *J. Nucl. Mater.*, 340 (1) (2005) 93–112.
[50] G. Leichtfried, Paper presented at Second International Conference on Tungsten and Refractory Metals, Metal Powder Industries Federation, Princeton, NJ, (1994,) 319.
[51] ITER-EDA, Final Design Report, Materials Assessment Report, 1998.
[52] S. M. Wolf, Properties and applications of dispersion-strengthened metals. *JOM* 19 (1967) 22–28.
[53] H. G. Sell, G. H. Keith, R. C. Koo, R. H. Schnitzel, R. Corth, Physical metallurgy of tungsten and tungsten base alloys, WADD-TR-60-37, Part II, Westinghouse Lamp Division, Bloomfield, NJ, (1961).
[54] J. Martinez, B. Savoini, M. A. Monge, A. Munoz, R. Pareja, Development of oxide dispersion strengthened W alloys produced by hot isostatic pressing, *Fusion Eng. Des.*, 86 (9–11) (2011) 2534–2537.
[55] P. M. Cheng, Z. J. Zhang, G. J. Zhang, J. Y. Zhang, K. Wu, G. Liu, W. Fu, J. Sun, Low cycle fatigue behaviors of pure Mo and Mo-La_2O_3 alloys, *Mater. Sci. Eng. A*, 707 (2017) 295–305.
[56] R. Bianco, R. W. Buckman, Mechanical properties of oxide dispersion strengthened (ODS) molybdenum, in: A. Crowson et al. (Eds.), *Molybdenum and Molybdenum Alloys*, TMS, San Antonio, Texas (1998) 125–144.

[57] J. B. Conway, P. N. Flagella, *Creep-Rupture Data for the Refractory Metals to High Temperatures*, Gordon and Breach Science Publishers, New York (1971) 576–678.

[58] A. J. Mueller, R. Bianco, R. W. Buckman, Evaluation of oxide dispersion strengthened (ODS) molybdenum and molybdenum–rhenium alloys, *Int. J. Refract. Met. Hard Mater.*, 18 (4–5) (2000) 205–211.

[59] B. V. Cockeram, The mechanical properties and fracture mechanisms of wrought low carbon arc cast (LCAC), molybdenum-0.5pct titanium-0.1pct zirconium (TZM), and oxide dispersion strengthened (ODS) molybdenum flat products. *Mater. Sci. Eng. A*, 418 (2006) 120–136.

[60] D. Sturm, M. Heilmaier, J. H. Schneibel, P. Jehanno, B. Skrotzki, H. Saageet, The influence of silicon on the strength and fracture toughness of molybdenum. *Mater. Sci. Eng.*, 463 (2007) 107–114.

[61] T. Zhang, H. W. Deng, Z. M. Xie, R. Liu, J. F. Yang, C. S. Liu, X. P. Wang, Q. F. Fang, Y. Xiong, Recent progress on designing and manufacturing of bulk refractory alloys with high performances based on controlling interfaces, *J. Mater. Sci. Technol.*, 52 (2020) 29–62.

[62] S. Wurster, R. Pippan, Nanostructured metals under irradiation, *Scripta Mater.*, 60 (2009) 1083–1087.

[63] Y. Bai, B. Dodd, *Adiabatic Shear Localization: Occurrence, Theories and Applications*, Pergamon Press, New York, (1992) 125–134.

[64] X. F. Liu, Z. L Tian, X. F. Zhang, H. H. Chen, T. W. Liu, Y. Chen, Y. J. Wang, L. H. Dai, "Self-sharpening" tungsten high-entropy alloy, *Acta Mater.*, 186 (2020) 257–266.

[65] R. Luo, D. Huang, M. Yang, E. Tang, M. Wang, L. He, Penetrating performance and "self-sharpening" behavior of fine-grained tungsten heavy alloy rod penetrators, *Mater. Sci. Eng. A*, 675 (2016) 262–270.

[66] S. Park, D. K. Kim, S. Lee, H. J. Ryu, S. H. Hong, Dynamic deformation behavior of an oxide-dispersed tungsten heavy alloy fabricated by mechanical alloying, *Metall. Mater. Trans A*, 32 (2001) 2011–2020.

3 Processing Fundamentals

3.1 CONVENTIONAL MELTING-CASTING

The melting-casting method is a well-established method of alloy fabrication. The high melting point of refractory metals is not economical for alloy development through melting, followed by casting, owing to the requirement for high temperature and oxygen entrapment. Even if the melting is carried out in an inert or vacuum environment, the major challenges still involve coarsening of the grain due to elevated temperature processing and less solubility of solute elements. Several methods such as: (1) electric-arc melting; (2) electron-beam melting; (3) electron beam zone refining; (4) cold crucibles-based induction melting; (5) levitation melting in a magnetic field; and (6) solar furnace-based melting in a vacuum are used to melt refractory metals and alloys, however, the process depends on the capacity of production. All the processes are not extensively used in industry and only a select few processing methods are described in this chapter. Ingots of Mo (diameter: 300 mm) of ≥ 1000 kg have been produced by melting [1]. Electric arc melting of Ta and Nb can be carried out without serious challenges and is generally adopted industrially to fabricate refractory alloys [1, 2]. It is also reported that vacuum arc melting leads to coarse columnar grain structure and the crystallite size is in millimeters (mm), which may prevent enhanced mechanical properties. Deoxidation also needs to be carried out to minimize the formation of porosity in the ingots, and the double remelting technique also needs to be carried out for an even distribution of the alloys and minimal impurity segregation [1]. Several techniques, such as rotating the crucible slag-based lining melting to refine the microstructure and enhance ductility are also implemented [3]. Li et al. have fabricated NbMoTa medium entropy alloy by the vacuum arc melting process and reported a dendritic substructure inside the grain [4]. Their paper also describes how vacuum arc melted alloy shows a compressive strength of 920 MPa at 1000°C [4]. Senkov et al. have fabricated $Nb_{25}Mo_{25}Ta_{25}W_{25}$ high entropy alloys by vacuum arc processing and observed a dendritic structure and segregation after solidification [5]. The highest stress and strain at the maximum stress of 1008 MPa and 16% of the alloy respectively at 1000°C have been reported [5]. The electron beam melting facilitates enhancement in refining refractory metals and improvement of ductility in Nb, Ta is reported [1]. The literature shows that a concentration of O, C, N, H significantly reduces in the vacuum-based electron beam melting process, compared to the argon or vacuum-based consumable arc melting processing of W, Ta, Nb [6]. In the electron beam melting (EBM) process, the metal powder is subjected to selective melting by an electron beam and a layered structure is formed. The beam diameter and location are regulated by two magnetic coils [7]. The EBM method is presented in Figure 3.1 [8]. The selective electron beam melting of pure W yields a density of 99.5%, compressive strength of 1560 MPa, a columnar

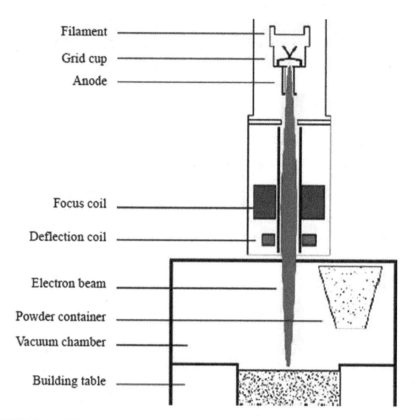

FIGURE 3.1 Schematic of EBM process. (Reprinted from [8]. Copyright (2016), with permission from Elsevier.)

microstructure, and the fracture occurs by grain boundary de-bonding and brittle transgranular mode [9]. Electron beam zone refining has been carried out to achieve high purity powder to enhance the final properties. Preparation of 99.999% pure W rods from Ammonium paratungstate through solvent extraction followed by electron beam floating-zone melting is discussed in the literature [10]. The extremely pure W is effective as a sputtering target in semiconductor, gate electrodes [10, 11]. The cold crucible-based melting technique is reported by Sterling and Warren, which involves induction melting of reactive metals in a non-reactive water-cooled crucible without creating any impurities [12]. Such a melting process leads to the formation of a solid layer which prevents direct contact between the metal to be melted and the crucible [13]. Several refractory metals such as Nb, Ta, Mo can be processed by cold crucible induction melting without any thermal shock for the crucible and melt contamination. The electromagnetic induction in the process causes the agitation and homogenization of the melt [14, 15]. Nb-Nb_3Si eutectic alloy has been developed by the cold crucible directional solidification method to prevent any contamination issues [15]. Ta-based sheet and plate fabrication mainly depends on the EBM process owing to its cost effectiveness [16]. Levitation melting assisted by a magnetic field is not a very

Processing Fundamentals

popular method, as very small amounts (a few grams) of melt can be achieved, though the melt is associated with better purity and homogenization [17]. Use of renewable solar energy in a furnace in a vacuum provides an economical approach to produce refractory oxides (at > 3000°C), however, because of the construction costs of a big enough solar furnace and the requirement of constant radiation, energy storage does not have the potential for wide applicability [18].

3.2 LIMITATIONS OF THE MELTING-CASTING ROUTE

The melting-casting processing route has several advantages, such as appreciable densification, elongation, and strength, however, at present, research on melting-casting-based fabrication of refractory metals and alloys is not extensive. Processing through the melting-casting route also depends on process viability and the cost factor. Reports indicate that only 2% Mo is produced through vacuum arc cast and electron beam melting [16]. Ta targets for the diffusion barrier in an integrated circuit are fabricated by ingot metallurgy or alternative powder metallurgy [16]. Alternatively, manufacturing of Ta wires for capacitor application requires powder metallurgy as that technique can provide the homogeneous microstructure and the necessary property [16]. High temperature processing of refractory metals requires a stringent vacuum or an inert atmosphere to prevent oxidation and also requires high energy for melting. Moreover, the coarsening of the grains at an elevated temperature for a long holding time reduces the strength of the materials. Near net shape processing is crucial to reduce the machining, improve the related processing cost and increase the yield, which is a real challenge in the melting-casting route. Fabrication of W filaments involves the drawing of W in wires and the ductility of the metal should be satisfactory. The as-cast structure of W is brittle and therefore involves challenges in processing. Variation in density, solubility and melting point during the processing of refractory based alloys through melting-casting [19] adds additional challenges, for example, the production of W-Ni-Fe (Tungsten heavy alloys) or W-Cu alloys. A large difference in density leads to segregation as the heavier element settles down under gravity during melting followed by casting [20]. A report illustrates that NbMoTa medium-entropy alloy fabricated by vacuum arc melting possesses coarser grain size (112.52 μm), enhanced segregation and the percentage strain at 1000°C is not enough (3.8%) for structural applications [4]. In the casting method, the solid solubility is not extensive and second-phase formation is coarser, which hinder the strengthening of the matrix by solid solution strengthening or the precipitation strengthening mechanism. Though several limitations of the melting-casting route have been reported, a couple of other studies indicate the potential of the process for structural applications [21, 22] and research is now directed toward refinement of the dendritic structure to improve plasticity [23].

3.3 MECHANICAL ALLOYING

Mechanical alloying (MA) is the room temperature, non-equilibrium alloy processing method to enhance the solid solubility, attain a refined grain size, a reduction in further processing temperature, and a near net shape component fabrication with superior properties. At present, mechanical alloying is a technique with great potential for designing alloys for critical applications. In the process, powders are subjected to stress-induced deformation by mechanical force. The powders undergo compression by grinding balls during the grinding in a vial and in wet or dry conditions. The grinding balls can be of WC-Co, tool steel, hardened steel, stainless steel, or tempered steel [24]. The end powder composition depends on the initial nature of the powders (ductile/brittle or both), the ball to powder weight ratio (BPR), the milling speed, the milling duration, the milling atmosphere, the process control agent, the temperature generated during milling, and the size of the grinding balls. The process results in minimal temperature evolution due to the friction effects, a decrease in particle size to nanometer level from initially micron-sized commercial powder particles. Due to grinding balls and powder contact, contamination from grinding balls can occur and adopting wet grinding (grinding in the presence of a process control agent such as stearic acid or toluene) minimizes the chances of contamination and cold welding by minimizing the surface tension of the powder particles [24]. The mechanical alloying can be carried out in various mills, such as SPEX shaker mills, planetary ball mills or attritor mills. Milling of 10–20 gm, 0.5–40 kg, 1250 kg can be milled in SPEX, attritor and commercial mills [24]. During the mechanical alloying of both ductile and brittle constituents, fracturing of brittle materials and flattening followed by strain hardening occur in ductile materials. The report suggests that alloying occurs if 15% of the initial material is ductile [25]. If the initial particles are ductile and brittle, the brittle constituents get encapsulated in the ductile material and further milling results in the uniform dispersion of the brittle particles. However, the alloying depends on the solid solubility of the brittle materials in the ductile material. Figure 3.2 shows that with the increase in milling time, some particles of Nb-Mo powder are flattened where other particles are refined after 1 h of milling [26]. The particles are progressively refined after 10 h of milling. The finer particles are difficult to fracture and tend to agglomerate, compared to the flattened larger particles which are refined. As a result, bimodal particle size distribution is achieved at the end of milling. The regulation of the particle size distribution is important in order to achieve superior packing and densification at the time of consolidation. The fine particle size significantly lowers the processing temperature, for example, around 0.4 time of melting temperature of W with appreciable density after heat treatment. The particle size refinement increases with an increase in milling energy, as superior refinement can be achieved in a high energy SPEX shaker mill compared to commercial mills.

The enhancement of solid solubility by mechanical alloying is attributed to the grain size refinement. The larger proportion of atoms in the grain boundary area results in a significant diffusion and alloying, assisted by the higher grain boundary energy due to the defect generation during mechanical alloying [24, 27]. Temperature generation during mechanical alloying can also contribute to atomic diffusion and solid solution. Mechanical alloying of Ni-based superalloys in attritor mills results in <100–215°C temperature rise [28]. Although temperature evolution facilitates alloying, it occurs mainly by elevated lattice strain and dislocation density with the progress of the mechanical alloying. A comparison of different non-equilibrium process routes with respect to solid solubility in Figure 3.3 shows that mechanical

Processing Fundamentals

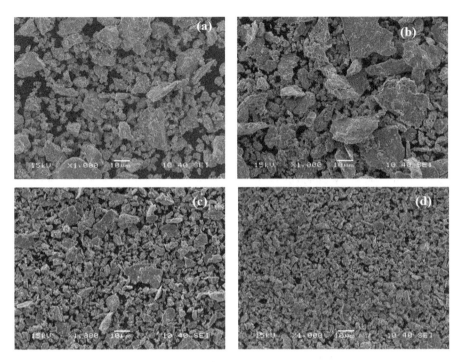

FIGURE 3.2 Scanning electron micrograph of Nb-Mo powder at (a) 0 h, (b) 1 h, (c) 5 h, (d) 10 h of mechanical alloying. (Reprinted from [26].)

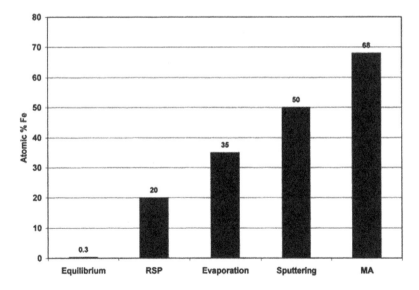

FIGURE 3.3 Solid solubility of Fe in Cu obtained by different non-equilibrium processing techniques. (Reprinted from [24]. Copyright (2001), with permission from Elsevier.)

alloying is effective to enhance the solid solubility [24], which is of considerable importance to improve the strength of the alloy.

3.4 RAPID SOLIDIFICATION

Processing of alloys through rapid solidification involves intense cooling (the cooling rate: 10^5–10^8 K/s) by increasing the heat extraction rate from a liquid metal and a significant departure from equilibrium [29]. The major benefits of the processing route are [30, 31]:

1. solid solubility enhancement;
2. grain size refinement;
3. homogeneous microstructure;
4. enhanced strength and decrease in anisotropy.

Faster solidification is possible using several techniques: high intensity of undercooling before solidification or an elevated cooling rate throughout solidification [32]. The rapid solidification process is classified into different categories.

First, melting spinning or melt extrusion to develop metallic glass and amorphous alloys as ribbons. Figure 3.4 presents the chill bock melt-spinning process where the liquid is made to fall on the rotating wheel via a nozzle to form a ribbon structure and the condition is maintained inert [33]. In this process, the width of the produced ribbon is only 3 mm, which is related to the maximum cooling rate and to avoid clogging of the nozzle during the melt flow. The anomalies lead to the development

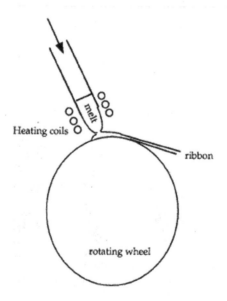

FIGURE 3.4 Chill block melt spinning apparatus. (Reprinted from [33]. Copyright (2000), with permission from Elsevier.)

of other methods such as planar flow casting and the melt overflow process [34]. The ribbon is formed by directing the melt into the cooled roll (speed of rotation: ~10 m/s) with a thickness of 10–50 μm [35]. The thickness (t) of the ribbon in the melt-spinning process can be expressed as [36]:

$$t = \left(\frac{\alpha_T}{V_R} \cdot \frac{\Delta T}{\Delta H} \right).l \qquad (3.1)$$

where α_T is empirical heat transfer co-efficient, $V_{R\,is}$ rotating wheel speed, ΔT is temperature variance at opposite ribbon side, ΔT is latent heat/unit volume of liquid metal, l is contact zone length among substrate and liquid metal.

Second, gas/water atomization by transferring liquid metal to aero-sol particles under high-pressure injection in a nozzle. The gas atomization process is illustrated in Figure 3.5. In the process, the liquid metal comes into contact with the inert gas and results in atomization due to kinetics energy transmission from an inert gas to a liquid metal [33]. The medium during the transformation can be nitrogen, argon or water. Very pure fine particles are produced by the process [37]. The size of the mean powder particle size grows with the increase in the pressure of the gas [38]. The capacity of the process varies between < 1 kg/run for laboratory use to 50,000 tonnes/year for commercial production [39].

Third, thermal spraying in which melt is speedily moved to the material on which it is solidified. This method is effective in fabricating composites, ceramics and alloys

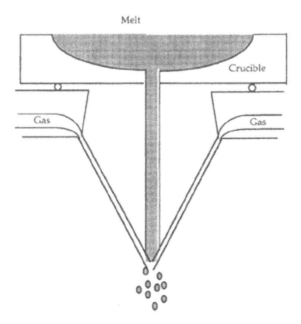

FIGURE 3.5 V- or cone-jet gas atomization apparatus. (Reprinted from [33]. Copyright (2000), with permission from Elsevier.)

with high temperature sustainability. Plasma spraying is a type of thermal spraying that provides ultrafine grains and the microstructure is metastable [40]. The microstructure obtained by the spray deposition technique depends on the liquid and solid fraction in the droplets, the temperature and the velocity of the deposition [41]. In the arc spraying method, the input is taken as wire (diameter: 0.16–0.32 cm) which melts to produce droplets [42, 43]. In this technique, the oxygen level is greatly reduced compared to plasma spraying, but the challenge in arc spraying is the suitability to apply alloyed wires [43, 44].

During rapid solidification, nucleation and growth of the crystal from the liquid metal are constrained and a non-equilibrium condition is achieved. Formation of the metastable phase with a lower melting temperature than the equilibrium liquid temperature is facilitated [36]. Yoshitake et al. have prepared refractory transition amorphous alloys by the melt-spinning technique using a Cu roller, and the outcome ribbon thickness is 10–20 μm, and a maximum crystallization temperature of 1456 K has been reported for $(Ta_{0.3}W_{0.7})_{80}Si_{10}B_{10}$ alloy [45]. Hyperperitectic Mo-40%Co alloy has been fabricated by vacuum arc melting (Ar atmosphere) and the electromagnetic levitation method [46]. It has been reported that the processing refines the grains, and a dendrite width of 3.01 μm has been achieved with undercooling of 252 K [46]. Application of Nb in aerospace requires oxidation-resistant coatings and minimum brittleness. Rapid solidification processing of Nb-based alloys forms amorphous oxides and the oxidation of the substrate is reduced [47]. Supercooling of Nb-35Ti, Nb-50Ti alloys (levitation melted) has been carried out at 200 K and 60–320 K respectively by the University of Florida [47]. It is anticipated that rapid solidification will be a promising prospect for the development of refractory alloys with enhanced solubility [47].

3.5 SEVERE PLASTIC DEFORMATION

Severe plastic deformation (SPD) is a cost-effective process to induce considerable strain without change in the area of the material and on the application of elevated hydrostatic pressure to inhibit crack propagation. Micron (mean grain size < ~1 μm) and nano-sized grains in a tube, a sheet, and several brittle materials such as amorphous glass can be fabricated by SPD [48, 49]. The fine grains formation by SPD occurs by a continuous dynamic recrystallization (high angle grain boundary formation) or a geometric dynamic recrystallization (formation of serrated grain boundary) process [49]. The literature indicates that for a 50% and a 90% decrease in height under uniaxial compression, the effective strains are 0.69 and 2.30, respectively and, for a plane strain compression, the effective strains for the identical decrease in height are 0.80 and 2.66, respectively [49]. The size of the subgrain (d) and dislocation density (ρ) is related as [49]:

$$d = \frac{\kappa}{\sqrt{\rho}} \tag{3.2}$$

The stress (σ) is also related to the dislocation density as:

$$\sigma = M\alpha\mu b \sqrt{\rho} \quad (3.3)$$

where M is the Taylor factor, α is the constant (0.5), μ is the shear modulus.

By replacing $\sqrt{\rho}$ from Equation (3.3) in Equation (3.2), it can be expressed as:

$$d = \frac{KM\alpha\mu b}{\sigma} \quad (3.4)$$

It is evident from Equation (3.4) that elevated stress during SPD results in finer grain size.

Researchers also indicate that microshear formation during SPD also leads to microstructural refinement [50]. During SPD, hydrostatic stresses and shear stress contribute to restrict cracking and the introduction of plastic strains respectively [49]. The SPD process can be carried out for the development of bulk product. Some of the SPD processes are discussed now.

3.5.1 Equal Channel Angular Pressing

In the equal channel angular pressing (ECAP) process, a sample is subjected to significant strain by introducing two ECAP die (channels) positioned at 90° by a plunger to produce ultrafine or nanostructured material without variation in the section area of the sample. The sample can be introduced into the die by various techniques as illustrated in Figure 3.6, which also influence the grain size reduction such as:

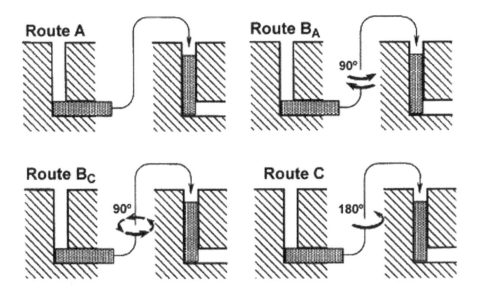

FIGURE 3.6 Four different fabrication methods in ECAP. (Reprinted from [51]. Copyright (2007), with permission from Elsevier.)

- (Route A) Identical (without changing direction) incorporation of sample during different passes.
- (Route B_A) Sample is rotated 90° about the longitudinal axis (identical direction in each pass).
- (Route B_C) Sample is rotated 90° (different direction in each pass).
- (Route C) Sample is rotated 180° about the longitudinal axis [51].

The processing of an Nb and Ta sample by indirect extrusion ECAP by rotating 90° in a clockwise direction with a decreased load, compared to conventional direct extrusion, is reported by Omranpour et al. [52]. Ultrafine W has also been produced by ECAP at temperatures of 800°C and 950°C [53]. Rubitschek et al. have fabricated refractory-based Nb-Zr alloy by ECAP (at room temperature) with rotation of 180°/90°/180° rotations about the longitudinal axis without any evidence of dynamic recovery [54].

3.5.2 Multidirectional Forging

Multidirectional forging (MDF) is a cost-effective method to fabricate large industrial products with the minimum cost of tools. In the process, a sample is deformed in a preheated lubricated die in several passes and the direction of loading is varied to 90° in an individual pass [55]. Figure 3.7 shows the MDF process where the sample is rotated between each pass as X→Y→Z→X and after completion of forging, the sample is cooled [55]. The MDF is limited in the processing of refractory metals, and several research studies have been carried out on the fabrication of Al, Mg, Ti, Cu, Zn alloys [56–62]. Bahmani et al. have fabricated Mg alloy by MDF at 220–340°C [59], whereas Zhang et al. have processed Cu Ni Si alloy through MDF at room temperature and in liquid nitrogen (-196°C) (cryogenic environment) and reported that cryogenic processing causes more refinement in the grain size compared to room temperature [61]. During MDF, the particle size distribution is enhanced, which

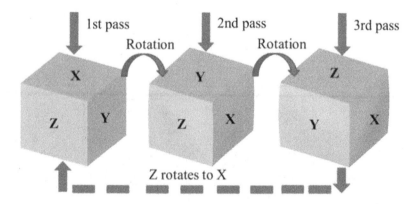

FIGURE 3.7 Schematic illustration of the MDF process. (Reprinted from [55]. Copyright (2016), with permission from Elsevier.)

contributes to the improvement of the mechanical properties [63]. Although MDF is advantageous to refine the grain structure, its effectiveness and recurrence [64], the grain size reduction is still less than that produced by other severe plastic deformation processes [65].

3.5.3 High Pressure Torsion

The high pressure torsion (HPT) process involves the addition of torsion-based shear strain through elevated hydrostatic pressure to develop nanomaterials [66, 67]. Research work conducted in 1988 on Al alloys provides a useful insight into the effectiveness of the HPT process for grain size reduction to submicron [68], however, the knowledge of such processing was developed in 1935 [69]. A disk-shaped sample of a low thickness is positioned between two anvils, followed by rotating the bottom anvil and increased pressure and shear strain are introduced for deformation at room temperature or elevated temperature [70, 71]. The HPT process and the factors used for assessing total strain are presented in Figure 3.8 and Figure 3.9 [72] and in the process, deformation of the disc occurs by quasi-hydrostatic pressure [71]. The dθ and dl in Figure 3.9 represent minor rotation and movement respectively. The true strain in the process can be calculated as:

$$\varepsilon = \ln\left(\frac{\varphi \cdot r}{h}\right) + \ln\left(\frac{h_0}{h}\right) \tag{3.5}$$

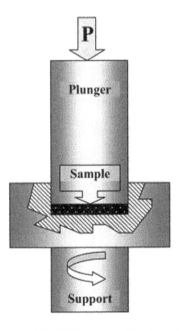

FIGURE 3.8 Schematic image of the HPT process. (Reprinted from [72]. Copyright (2003), with permission from Elsevier.)

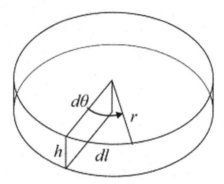

FIGURE 3.9 Required parameters for assessing the total strain in HPT. (Reprinted from [72]. Copyright (2003), with permission from Elsevier.)

as, $(\varphi \cdot r)/h \gg 1$ and $\varphi = 2\pi N$

$$\varepsilon = \ln\left(\frac{2\pi N.r.h_0}{h^2}\right) \quad (3.6)$$

where h_0 is the initial thickness of the sample, h is the final thickness of the sample, r is the disk radius, N is the revolutions number.

The HPT process can be classified into constrained and unconstrained. Constrained HPT involves positioning the sample inside the hollow section of the bottom anvil and the outside movement of the sample is restricted during the application of torsional strain and back pressure is applied, though in the actual experiment the mechanism is quasi-constrained. Unconstrained HPT includes non-hindrance during movement of the sample and the sample is positioned on the bottom anvil rather than inside the section of the anvil and nominal back-pressure is included [71, 72]. It is also reported that HPT is capable of producing homogeneous and equiaxed microstructures [73]. Several research studies on the deformation of refractory metals and alloys by HPT have been carried out. Li et al. have fabricated disk-based Mo (diameter: 10 mm, thickness: 1 mm) by HPT in a quasi-constrained state at 623 K and pressure application of 3 GPa, speed of rotation of 0.67 rpm, 5, 10, 15 number of revolutions were maintained during the study [74]. Hf, Hf-Ti alloys have been deformed at room temperature by HPT with 4.5 GPa pressure, rotational speed of 0.3 rpm and it has been reported that an HPT-processed product possesses more refinement in grain size and homogeneity compared to a cast product [75]. Deformation of Ta by the HPT technique occurred at room temperature by application of 6.0 GPa pressure with the bottom anvil speed of rotation kept at 1 rpm [76].

3.5.4 ACCUMULATIVE ROLL BONDING

The accumulative roll bonding (ARB) process was established in 1998 [77] where both rolling and bonding are integrated to attain high strength of large products

Processing Fundamentals

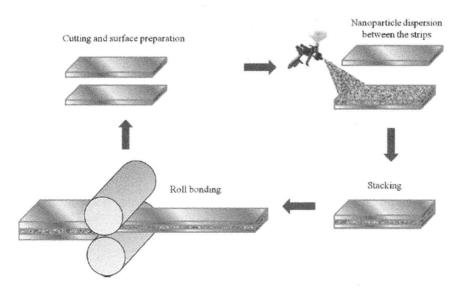

FIGURE 3.10 Schematic illustration of the accumulative roll bonding (ARB) method. (Reprinted from [78]. Copyright (2018), with permission from Elsevier.)

without variation in the dimensions. In the process, sheets are rolled to reduce thickness (50%), and then are cut into two pieces which are subjected to surface treatment (degreasing, wire brushing) and piled upon each other followed by rolling. Fabrication of nano-composites was carried out by ARB through the dispersal of nanoparticles between two sheets followed by rolling of the sheets [78]. The ultimate properties depend on the number of passes and are decided by the bonding efficiency and evenness in particle dispersion [78]. The temperature of the process should be below the recrystallization temperature to enhance the bonding and decrease the rolling load [79]. Figure 3.10 illustrates the ARB process. The equivalent strain (ε) as measured by von Mises as related to the ARB process and can be presented as [80]:

$$\varepsilon = \left\{ \frac{2}{\sqrt{3}} \ln\left(\frac{1}{2}\right) \right\} \times n = 0.80n \qquad (3.7)$$

where n is the number of cycles.

Adequate surface treatment is required for efficient roll bonding, however, during the ARB process, certain defects of the sheet, such as fracture or edge cracking can occur, which can be minimized by proper regulation of the processing parameters in each cycle [79]. The microstructure in the ARB process depends on the applied strain and it is reported that homogeneous ultra-fine grain can be obtained when the cumulative equivalent strain is > 5 [79]. The advanced ARB process, denoted as accumulative roll bonding and folding (ARBF), has been developed where roll bonding and folding are included and therefore the cutting operation is eliminated [81].

3.5.5 Twist Extrusion

In the twist extrusion (TE) process, a die with a twist section as presented in Figure 3.11 is used for extrusion of the sample by applying forward and backward pressure through dummy blocks [82]. The TE process produces long and complex products, and hollow, rectangular portions can be extruded by the process [82, 83]. The TE process is advantageous compared to HPT in order to increase the sample size [84]. The deformation during TE is homogeneous and isotropic through the application of a clockwise–anticlockwise–clockwise sequence of twisting [83]. During TE, the length: diameter of the sample needs to be less so that the plunger during application pressure does not undergo bending and the plunger traverses the restricted distance [84]. The sample in the TE process is subjected to twisting and untwisting, but the final sample sizes do not differ with respect to the initial size [83, 85]. Formation of ultrafine grains for pure Ti size assisted by the dislocation slip in the TE process is reported in the literature [86].

The stresses induced in the die are $(\sigma) \propto \varepsilon_0$, where ε_0 is the von Mises strain/pass [87]. The sustainability of the die in terms of passes (N) is proportional to $(\varepsilon_0)^{-2}$. The pass (n) to incorporate the von Mises strain $\varepsilon_{(n)}$ is inversely proportional to ε_0. Therefore, the quantity of samples fabricated by TE retaining the sustainability of the die is [87]:

$$\frac{N}{n} = \frac{1}{\varepsilon_0} \quad (3.8)$$

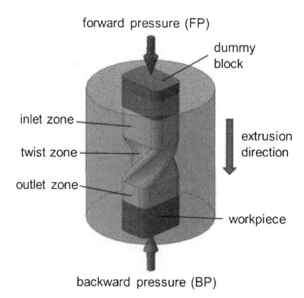

FIGURE 3.11 Twist extrusion method. (Reprinted from [82]. Copyright (2012), with permission from Elsevier.)

3.6 THE SOL-GEL METHOD

The sol-gel process includes forming three-dimensional structures by a polymerization reaction initiated with ions, inorganic metal salts, or metal-organic compounds [88]. The method can produce ultrafine structured particles, nanoporous membranes and nanofilms [88]. Hydrolysis and condensation are involved in the process [89]. During hydrolysis, an unstable M-OH bond develops and throughout the condensation phase the M-OH or M-OR (if the initial material is alkoxides) combines to give rise to an M-O-M bond. The reactions are provided below [89]:

$$M(OR)_n + H_2O \rightarrow M(OR)_{n-1}(OH) + ROH \text{ (during hydrolysis)}$$

$$M(OR)_{n-1}(OH) + (HO)M(OR)_{n-1} \rightarrow M(OR)_{n-1}\text{M-O-M}(OR)_{n-1} + H_2O \text{ (during condensation)}$$

or,

$$M(OR)_{n-1}(OH) + M(OR)_n \rightarrow M(OR)_{n-1}\text{M-O-M}(OR)_{n-1} + ROH \text{ (during condensation)}$$

The processes result in the formation of suspended dispersed particles denoted as sol. The chemical processing of sol develops attached particles along with the liquid present in the pores [89]. Furthermore, evaporation of the liquid develops a nanoporous/dense film structure [88]. Sol-gel is applied for the development of stoichiometric, equiaxed, uniform and dense structures. Titanium triethanolamine (TEOA) is subjected to hydrolysis to produce Ti particles (5–30 nm) and the particle size and shape are regulated by the pH factor [90–92]. Reports also indicate that metal/ceramic nanocomposites can be prepared with regulated heterogeneity at the nanoscale [93]. The sol-gel method has several benefits, such as: (1) regulated reaction kinetics; (2) simplicity of alteration of the composition; (3) microstructure tailoring; and (4) the addition of different functional assemblies [94–96]. Liu et al. have prepared an W-Y_2O_3 alloy using the sol-gel method through the chemical reaction of ammonium paratungstate $(NH_4)_{10}[H_2W_{12}O_{42}]\cdot 4H_2O$ and yttrium nitrate hydrate $Y(NO_3)_3\cdot 6H_2O$ followed by hydrogen-based reduction of the formed oxides of W and yttrium (Y) [97]. It is reported that the particle size of W formed at the 800°C reduction temperature is 100–400 nm [97]. The paper also reports that a higher reduction temperature (850°C) enhances the particle size of W to 0.2–0.3 μm and the distribution of nano-Y_2O_3 inside the W matrix is homogeneous [97]. Ting et al. have suggested that the sol-gel and the calcination process can be modified to sol-gel and a mechanochemical technique for an appreciable reduction in the particle size during the synthesis of SiO_2–TiO_2 nanocomposites [98]. In comparing sol-gel and mechanical alloying, the literature indicates that mechanical alloying is a beneficial powder synthesis method [99].

3.7 CHEMICAL VAPOR DEPOSITION

Chemical vapor deposition (CVD) is a method to produce two-dimensional (2D) nanostructures, however, nanowires and powder can be synthesized [100, 101]. In

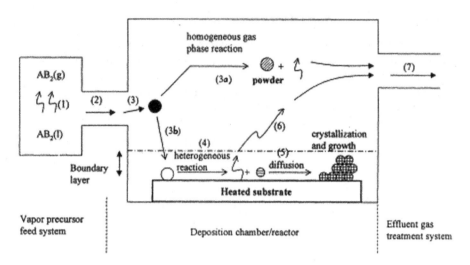

FIGURE 3.12 Stages in main CVD stages in deposition. (Reprinted from [102]. Copyright (2003), with permission from Elsevier.)

the method, precursors (hydrides, chlorides, carbonyls, materials with high tendency to sublimation) can be subjected to sublimation and dissociate at the heated substrate to produce solid materials [101]. The gas phase-based reaction results in the formation of fine particles and the process can also be regulated to produce thin films [100, 101]. Figure 3.12 presents the CVD process involving a precursor feed section for the production of the vapor phase and the transfer to the deposition chamber, a deposition chamber for substrate heating and reaction, and an effluent gas treatment section [100, 102]. Homogeneous reaction in the deposition chamber produces particles, but both homogeneous and heterogeneous reactions can occur to develop porous films [100, 102]. Depending on the heating method, CVD can be categorized into thermal CVD, plasma CVD, laser-facilitated CVD, and ion-aided CVD [100]. The deposition rate (k) is related to the deposition temperature (T) as [102]:

$$k = A\exp\left(\frac{-E_a}{RT}\right) \qquad (3.9)$$

where A is the constant, E_a is the apparent activation energy, R is the gas constant.

The growth of the deposited layer is interface-regulated at a low temperature and transport-regulated at an elevated temperature. The deposited thickness depends on the rate of deposition. A low deposition rate produces epitaxial thin films used in the semiconductor sector and a high deposition rate results in coating with increased thickness [102].

The literature reports the synthesis of W from the dissociation of WCl_6 or WF_6 by H_2 [102]. The temperature of the deposition for WF_6 is 550°C compared to almost 1000°C for WCl_6 and therefore it is applied as an effective precursor [102]. Though

CVD is useful to synthesize materials with high density and purity, the formation of hazardous gas as a precursor, using precursors from more than one source, limits the use of the process [102].

3.8 PHYSICAL VAPOR DEPOSITION

In the physical vapor deposition (PVD) method, material is subjected to the evaporation of material (solid/liquid) followed by transfer of the formed vapor in a vacuum/reduced pressure-based chamber and condensation of the substrate [103]. Figure 3.13 shows the PVD process involving the evaporation and ion beam or plasma-based sputtering on the substrate [104]. For evaporation in a vacuum the pressure requirement is 10^{-5} torr – 10^{-9} torr, and the rate of free surface vaporization of 10^{-3} g/cm^2/sec is required for effective deposition [105]. The vaporization can be in atomic, atomic clusters or molecules (with/without dissociation) [105]. In the process, the coating can be consistent, the coating thickness can vary < 1 nm–200 μm, and also multilayer coating can be done [106]. The rate of deposition in the process varies between 1–10 nm/s and application of the process ranges from small sample dimensions to large (10' ×12' size glass panels) which can be produced by the process [103]. In a comparison between the CVD process (chemical reaction-based deposition) and PVD (physical process-based deposition), the PVD is useful for a wide variety of materials [106]. Several deposition methods have been investigated, such as magnetron sputter deposition (MSD), reactive cathodic arc deposition (CAD) and high power impulse magnetron sputtering (HiPIMS) in the development of high entropy alloy (MoNbTaVW) thin films [107]. The paper reports that growth due to high energy

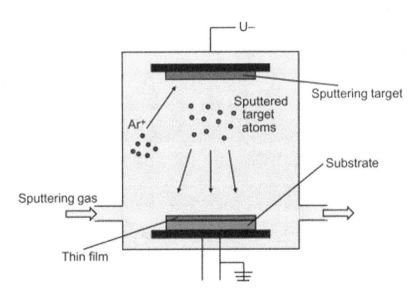

FIGURE 3.13 PVD technique. (Reprinted from [104]. Copyright (2018), with permission from Elsevier.)

in CAD and HiPIMS assists in the production of high density, compressive strength and the hardness of the film [107]. The PVD process also has certain limitations such as increased cost, low yield, and challenges in coating complex structures [106].

3.9 LASER ABLATION

Laser ablation is a potential method to produce nanoparticles, nanorods, nanocubes, or nanocomposites in a liquid, vacuum or gaseous atmosphere [108]. In this method, a laser penetrates the sample and the electron from the sample is removed by the intense electric field. The interaction between the electron and the material's atoms leads to surface heating by energy transition followed by vaporization. The formed plasma by the intense laser impingement on the surface undergoes expansion and cooling due to the pressure variance between the atmosphere and the plasma [108, 109]. During plasma condensation, the development of a nanoscale structure occurs. The depth of the laser penetration (10 nm) is related to the wavelength of the laser and the refraction index of the material [108]. Mikhailov et al. have developed a refractory metal thin film through the laser ablation deposition method with a power density of almost 30 J/cm^2/pulse, using an AY:Nd^{3+} laser and a vacuum level in the chamber of ≥ 10^{-9} torr [110]. The literature indicates the film thickness is 60 nm and 80 nm for Nb and W respectively [110]. Coalescence of the particles generated by laser ablation is a concern and, therefore, the process is carried out at low gas pressure conditions to promote the particle size refinement and decrease the tendency of agglomeration [111]. Kawakami et al. have fabricated nanostructured W particles by the laser irradiation method using Nd:YAG lasers, He gas, at various laser powers and gas pressure and reported that enhancement of the laser power and gas pressure increases the particle size of W [112].

3.10 INFERENCES

This chapter discusses various nanomaterials synthesis techniques, including the advantages, the regulating process parameters and the challenges. The conventional melting casting processing method possesses challenges with respect to process economy, grain size control and attainment of the desired final properties. Non-equilibrium synthesis techniques such as mechanical alloying, rapid solidification and other deformation-assisted methods, such as severe plastic deformation, chemical method (sol-gel), vapor and laser-assisted synthesis, offer advantages to refine the final microstructure. Mechanical alloying as a solid-state synthesis method is effective to synthesize ODS superalloys. Additionally, immiscible elements can lead to alloy formation by mechanical alloying [24]. In the early years of mechanical alloying, between 1970 to 1994, there was a significant number of publications [24]. Mechanical alloying has significant potential to develop nanostructured high entropy alloys with increased phase stability of the solid solution, and the number of publications between 2016–2018 is 130 [113]. The improvement in the diffusion of mechanical alloying due to nanostructure formation is an essential aspect of forming a solid solution. Mechanical alloying in a planetary ball mill or attritor mills is also an efficient way to produce commercial quantities of the product in contrast to the deposition or chemical methods [113].

REFERENCES

[1] E. M. Savitskii, G. S. Burkhanov, Melting and treatment of refractory metals and alloys, in: E. M. Savitskii, G. S. Burkhanov (Eds.), *Physical Metallurgy of Refractory Metals and Alloys*, Springer, Boston, MA, (1970) 235–283.

[2] V. I. Dobatkin, N. F. Anoshkin, A. L. Andreev, et al., *Ingots of Titanium Alloys*, Metallurgiya, Moscow, (1966) [in Russian].

[3] V. L. Girshov, L. N. Postnov, *Vacuum Metallurgy*, Izd. Doma Nauchno-Tekhn. Propagandy, Leningrad, (1964) [in Russian].

[4] Q. Li, H. Zhang, D. Li, Z. Chen, F. Wang, M. Wu, Comparative study of the microstructures and mechanical properties of laser metal deposited and vacuum arc melted refractory NbMoTa medium-entropy alloy, *Int. J. Refract. Met. Hard Mater.*, 88 (2020) 105195.

[5] O. N. Senkov, G. B. Wilks, J. M. Scott, D. B. Miracle, Mechanical properties of $Nb_{25}Mo_{25}Ta_{25}W_{25}$ and $V_{20}Nb_{20}Mo_{20}Ta_{20}W_{20}$ refractory high entropy alloys, *Intermetallics*, 19 (2011) 698–706.

[6] T. E. Butler, R. P. Morgan, ARC melting in high-vacuum environments, *JOM*, 14 (3) (1962) 200–203.

[7] H. Galarraga, D. A. Lados, R. R. Dehoff, M. M. Kirka, P. Nandwana, Effects of the microstructure and porosity on properties of Ti-6Al-4V ELI alloy fabricated by electron beam melting (EBM), *Addit. Manuf.*, 10 (2016) 47–57.

[8] R. Singh, S. Singh, M. S. J. Hashmi, Implant materials and their processing technologies, in: S. Hashmi, S. Mridha (Eds.), *Reference Module in Materials Science and Materials Engineering*, Elsevier, Oxford, (2016) 1–31.

[9] G. Yang, P. Yang, K. Yang, N. Liu, L. Jia, J. Wang, H. Tang, Effect of processing parameters on the density, microstructure and strength of pure tungsten fabricated by selective electron beam melting, *Int. J. Refract. Met. Hard Mater.*, 84 (2019) 105040.

[10] H. S. Yu, J. S. Kim, K. I. Rhee, J.-C. Lee, H. Y. Sohn, Production of ultrapure tungsten by solvent extraction with a D2EHPA/TOPO mixture and electron-beam zone refining, *Int. J. Refract. Met. Hard Mater.*, 11(5) (1992) 317–324.

[11] J. Y. Chen, Introduction, in: E. K. Broadbent (Ed.), *Tungsten and Other Refractory Metals for VLSI Applications*, II, Materials Research Society, Pittsburgh, PA, (1986) 3–6.

[12] H. F. Sterling, R. W. Warren: High temperature melting without contamination in cold crucibles, *Metallurgia*, 67 (1963) 301–307.

[13] A. Mühlbauer, Innovative induction melting technologies: A historical review, paper presented at International Scientific Colloquium Modelling for Material Processing, Riga, June 8–9, 2006.

[14] K. Sakamoto, K. Yoshikawa, T. Kusamichi, T. Onoye, Changes in oxygen contents of titanium aluminides induction, cold crucible induction and electron beam by vacuum melting, *ISIJ International*, 32 (5) (1992) 616–624.

[15] K. M. Chang, B. P. Bewlay, J. A. Sutliff, M. R. Jackson, Cold-crucible directional solidification of refractory metal-silicide eutectics, *JOM*, 44 (1992) 59–63.

[16] J. R. Ciulik, Jr. J. A. Shields, P. Kumar, T. Leonhardt, J. L. Johnson, Properties and selection of powder metallurgy refractory metals, in: P. Samal, J. Newkirk (Eds.), *Powder Metallurgy*, vol. 7, *ASM Handbook*, ASM International, Materials Park, OH, (2015) 593–598.

[17] N. H. El-kaddah, D. C. C. Robertson, The kinetics of gas-liquid metal reactions involving levitated drops, *Metall. Trans.*, 9B (1978) 191–199.
[18] J. C. Mcveigh, Thermal power and other thermal applications, in: J. C. Mcveigh (Ed.), *Sun Power*, (2nd edn), Pergamon, Oxford, (1983) 132–160,
[19] R. Morales, R. E. Aune, S. Seetharaman, O. Grinder, The powder metallurgy processing of refractory metals and alloys, *JOM*, 55 (2003) 20–23.
[20] G. Z. Chen, Forming metal powders by electrolysis, in: I. Chang, Y. Zhao (Eds.), *Advances in Powder Metallurgy: Properties, Processing and Applications*, Woodhead Publishing Limited, Cambridge, (2013) 33.
[21] J. Zhang, Y. Hu, Q. Wei, Y. Xiao, P. Chen, G. Luo, Q. Shen, Microstructure and mechanical properties of RexNbMoTaW high-entropy alloys prepared by arc melting using metal powders, *J. Alloys Compd.*, 827 (2020) 154301.
[22] M. Wang, R. Li, T. Yuan, C. Chen, M. Zhang, Q. Weng, J. Yu, Selective laser melting of W-Ni-Cu composite powder: Densification, microstructure evolution and nano-crystalline formation, *Int. J. Refract. Met. Hard Mater.*, 70 (2018) 9–18.
[23] X. J. Gao, L. Wang, N. N. Guo, L. S. Luo, G. M. Zhu, C. C. Shi, Y. Q. Su, J. J. Guo, Microstructure characteristics and mechanical properties of Hf0.5Mo0.5NbTiZr refractory high entropy alloy with Cr addition, *Int. J. Refract. Met. Hard Mater.*, 95 (2021) 105405.
[24] C. Suryanarayana, Mechanical alloying and milling, *Prog. Mater. Sci.*, 46 (1–2) (2001) 1–184.
[25] J. S. Benjamin, Mechanical alloying, *Sci. Amer.*, 234 (5) (1976) 40–49.
[26] D. K. Gardia, A. Patra, Synthesis and characterization of Nb-W and Nb-Mo alloys for high temperature structural applications, B.Tech dissertation, National Institute of Technology Rourkela, (2019).
[27] G. Veltl, B. Scholz, H-D. Kunze, Amorphization of Cu-Ta alloys by mechanical alloying, *Mater. Sci. and Eng. A*, 134 (1991) 1410–1413.
[28] A. B. Borzov, E. Ya Kaputkin, Method for measurement of temperature in attritor during mechanical alloying, in: J. J. deBarbadillo, et al. (Eds.), *Mechanical Alloying for Structural Applications*, ASM International, Materials Park, OH (1993) S1–S4.
[29] D. Turnbull, Metastable structures in metallurgy, *Metall. Trans.*, 12A (1981) 695–708.
[30] T. Dorin, A. Vahid, J. Lamb, Aluminium lithium alloys, in: R. N. Lumley (Ed.), *Fundamentals of Aluminium Metallurgy*, Woodhead Publishing, Cambridge, (2018) 387–438.
[31] B. Badan, L. Giordano, E. Ramous, S. Rudilosso, Rapid solidification of surface layers on alloy steels melted by CW laser, in: S. Steeb, H. Warlimont (Eds.), *Rapidly Quenched Metals*, Elsevier, Oxford, (1985) 851–854.
[32] R. E. Lewis, I. G. Palmer, J. C. Ekvall, I. F. Sakata, W. E. Quist, Aerospace structural applications of rapidly solidified aluminum-lithium alloys, in: *Proceedings of the Third International Conference on Rapid Solidification Processing, Gaithersburg, MD, December 1982*, National Bureau of Standards, Washington, DC, (1983) 613.
[33] I. T. H. Chang, Rapid solidification processing of nanocrystalline metallic alloys, in: H. S. Nalwa (Ed.), *Handbook of Nanostructured Materials and Nanotechnology*, Academic Press, New York, vol. 1, (2000) 501–532.
[34] E. J. Lavernia, T. S. Srivatsan, The rapid solidification processing of materials: Science, principles, technology, advances, and applications, *J. Mater. Sci.*, 45 (2010) 287–325.
[35] H. Jones, Rapid solidification, in: C. Suryanarayana (Ed.), *Pergamon Materials*, 2, Pergamon, Oxford, (1999) 23–45.
[36] T. R. Anthony, H. E. Cline, Dimensional variations in Newtonian-quenched metal ribbons formed by melt spinning and melt extraction, *J. Appl. Phys.*, 50 (1979) 245.

[37] M. Sherif El-Eskandarany, Introduction, in: M. Sherif El-Eskandarany (Ed.), *Mechanical Alloying* (3rd edn), William Andrew Publishing, Norwich, (2020) 1–11.
[38] J. S. Thompson, A study of process variables in the production of aluminum powder, *J. Inst. Met.*, 7 (1948) 101.
[39] Anon, *Adv. Mater. Process.*, 135 (6) (1989) 12.
[40] S. Sampath, H. Herman, Plasma spray forming metals, intermetallics, and composites, *JOM*, 45 (1993) 42–49.
[41] T. Egami, W. L. Johnson, Introduction and background, in: M. A. Otooni (Ed.), *Elements of Rapid Solidification Fundamentals and Applications*, Springer, Berlin, 29 (2013) 1–20.
[42] I. E. Locci, R. D. Noebe, The role of rapid solidification processing in the fabrication of fiber reinforced metal matrix composites, *NASA Technical Memorandum*, NASA-TM-101450, (1989) 1–45.
[43] L. J. Westfall, D. W. Petrasek, D. L. McDanels, T. L. Grobstein, Preliminary feasibility studies of tungsten/Niobium composites for advanced power systems applications, *NASA Technical Memorandum*, NASA TM 87248, (1986).
[44] E. Erturk, H. D. Steffens, Low pressure arc spraying in comparison with low pressure plasma spraying, in: *Advances in Thermal Spraying*, *Proc. of the 11th Int. Thermal Spray Conf.*, Pergamon Press, New York, (1987) 83.
[45] T. Yoshitake, Y. Kubo, H. Igarashi, Preparation of refractory transition metal-metalloid amorphous alloys and their thermal stability, *Mater. Sci. Eng.*, 97 (1988) 269–271.
[46] S. Sha, W. L. Wang, J. Chang, Y. H. Wu, S. S. Xu, B. Wei, Rapid growth kinetics and microstructure modulation of intermetallic compounds within containerlessly solidifying Mo-Co refractory alloys, *Intermetallics*, 125 (2020) 106886.
[47] J. Wadsworth, T. G. Nieh, J. J. Stephens, Recent advances in aerospace refractory metal alloys. *Int. Mater. Rev.*, 33 (1) (1988) 131–150.
[48] G. Faraji, H. S. Kim, H. T. Kashi, Fundamentals of severe plastic deformation, in: G. Faraji, H. S. Kim, H. T. Kashi (Eds,), *Severe Plastic Deformation*, Elsevier, Oxford, (2018) 19–36.
[49] R. Kapoor, Severe plastic deformation of materials, in: A. K. Tyagi, S. Banerjee (Eds.), *Materials Under Extreme Conditions*, Elsevier, Oxford, (2017) 717–754.
[50] T. Sakai, A. Belyakov, H. Miura, Ultrafine grain formation in ferritic stainless steel during severe plastic deformation, *Metall. Mater. Trans. A*, 39 (2008) 2206–2214.
[51] T. G. Langdon, The principles of grain refinement in equal-channel angular pressing, *Mater. Sci. Eng. A*, 462 (1–2) (2007) 3–11.
[52] B. Omranpour, L. Kommel, V. Mikli, E. Garcia, J. Huot, Nanostructure development in refractory metals: ECAP processing of Niobium and Tantalum using indirect-extrusion technique, *Int. J. Refract. Met. Hard Mater.*, 79 (2019) 1–9.
[53] T. Hao, Z. Q. Fan, T. Zhang, G. N. Luo, X. P. Wang, C. S. Liu, Q. F. Fang, Strength and ductility improvement of ultrafine-grained tungsten produced by equal-channel angular pressing, *J. Nucl. Mater.*, 455 (2014) 595–599.
[54] F. Rubitschek, T. Niendorf, I. Karaman, H. J. Maier, Microstructural stability of ultrafine-grained niobium–zirconium alloy at elevated temperatures, *J. Alloys Compd.*, 517 (2012) 61–68.
[55] H. Huang, J. Zhang, Microstructure and mechanical properties of AZ31 magnesium alloy processed by multi-directional forging at different temperatures, *Mater. Sci. Eng. A*, 674 (2016) 52–58.
[56] M. S. Kishchik, A. D. Kotov, D. O Demin, A. A. Kishchik, S. A. Aksenov, A. V. Mikhaylovskaya, The effect of multidirectional forging on the deformation and microstructure of the Al–Mg Alloy. *Phys. Metals Metallogr.*, 121 (2020) 597–603.

[57] S. Nouri, M. Kazeminezhad, A. Shadkam, Flow stress of 2024 aluminum alloy during multi-directional forging process and natural aging after plastic deformation, *Mater. Chem. Phy.*, 254 (2020) 123446.

[58] J. Wei, S. Jiang, Z. Chen, C. Liu, Increasing strength and ductility of a Mg-9Al alloy by dynamic precipitation assisted grain refinement during multi-directional forging, *Mater. Sci. Eng. A*, 780 (2020) 139192.

[59] A. Bahmani, S. Arthanari, K. S. Shin, Achieving a high corrosion-resistant and high strength magnesium alloy using multi directional forging, *J. Alloys Compd.*, 856 (2021) 158077.

[60] Z. X. Zhang, S. J. Qu, A. H. Feng, J. Shen, Achieving grain refinement and enhanced mechanical properties in Ti–6Al–4V alloy produced by multidirectional isothermal forging, *Mater. Sci. Eng. A*, 692 (2017) 127–138.

[61] R, Zhang, Z. Li, X. Sheng, Y. Gao, Q. Lei, Grain refinement and mechanical properties improvements in a high strength Cu–Ni–Si alloy during multidirectional forging, *Fusion Eng. Des.*, 159 (2020) 111766.

[62] P. C. Sharath, K. R. Udupa, G.V.P. Kumar, Effect of multi directional forging on the microstructure and mechanical properties of Zn-24 wt% Al-2 wt% Cu Alloy. *Trans. Indian Inst. Met.*, 70 (2017) 89–96.

[63] C. Selcuk, A. R. Kennedy, Al–TiC composite made by the addition of master alloys pellets synthesised from reacted elemental powders, *Mater. Lett.*, 60 (2006) 3364–3366.

[64] L. Tang, C. Liu, Z. Chen, D. Ji, H. Xiao, Microstructures and tensile properties of Mg-Gd-Y-Zr alloy during multidirectional forging at 773K, *Mater. Des.*, 50 (2013) 587–596.

[65] F. Djavanroodi, M. Ebrahimi, J. Nayfeh, Tribological and mechanical investigation of multi-directional forged nickel, *Sci. Rep.*, 9 (2019) 241.

[66] R. Z. Valiev, Y. Estrin, Z. Horita, T. G. Langdon, M. J. Zehetbauer, Y. T. Zhu, Producing bulk ultrafine-grained materials by severe plastic deformation, *JOM*, 58 (4) (2006) 33–39.

[67] R. Wadsack, R. Pippan, B. Schedler, Structural refinement of chromium by severe plastic deformation, *Fusion Eng. Des.*, 66–68 (2003) 265–269.

[68] Z. Horita, Production of multifunctional materials using severe plastic deformation, in: *Proceedings of the International Symposium on Giant Straining Process for Advanced Materials*, Kyushu University Press, Fukuoka, (2010).

[69] P. W. Bridgman, Effects of high shearing stress combined with high hydrostatic pressure, *Phys. Rev.*, 48 (1935) 825–847.

[70] K. Edalati, A. Yamamoto, Z. Horita, T. Ishihara, High-pressure torsion of pure magnesium: Evolution of mechanical properties, microstructures and hydrogen storage capacity with equivalent strain, *Scripta Mater.*, 64 (2011) 880–883.

[71] A. P. Zhilyaev, T. G. Langdon, Using high-pressure torsion for metal processing: Fundamentals and applications, *Prog. Mater. Sci.*, 53 (6) (2008) 893–979.

[72] A. P. Zhilyaev, G. V. Nurislamova, B. K. Kim, M. D. Baro, J. A. Szpunar, T. G. Langdon, Experimental parameters influencing grain refinement and microstructural evolution during high-pressure torsion, *Acta Mater*, 51 (2003) 753–765.

[73] A. P. Zhilyaev, T. R. McNelley, T. G. Langdon, Evolution of microstructure and microtexture in fcc metals during high-pressure torsion, *J. Mater. Sci.*, 42 (2007) 1517–1528.

[74] P. Li, Q. Lin, X. Wang, Y. Tian, K.-M. Xue, Recrystallization behavior of pure molybdenum powder processed by high-pressure torsion, *Int. J. Refract. Met. Hard Mater.*, 72 (2018) 367–372.

[75] A. V. Dobromyslov, N. I. Taluts, V. P. Pilyugin, Severe plastic deformation by high-pressure torsion of Hf and Hf single bond Ti alloys, *Int. J. Refract. Met. Hard Mater.*, 93 (2020) 105354.

[76] N. Maury, N. X. Zhang, Y. Huang, A. P. Zhilyaev, T. G. Langdon, A critical examination of pure tantalum processed by high-pressure torsion, *Mater. Sci. Eng. A*, 638 (2015) 174–182.
[77] Y. Saito, N. Tsuji, H. Utsunomiya, T. Sakai, R. G. Hong, Ultra-fine grained bulk aluminum produced by accumulative roll-bonding (ARB) process, *Scr. Mater.*, 39 (1998) 1221–1227.
[78] C. M. C. Jimenez, M. T. P. Prado, Processing of nanoparticulate metal matrix composites, in: P. W. R. Beaumont, C. H. Zweben (Eds.), *Comprehensive Composite Materials*, vol. IV, Elsevier, Oxford, (2018) 313–330.
[79] N. Tsuji, Accumulative roll-bonding, in: K. H. Jürgen Buschow, R. W. Cahn, M. C. Flemings, B. Ilschner, E. J. Kramer, S. Mahajan, P. Veyssière (Eds.), *Encyclopedia of Materials: Science and Technology*, Elsevier, Oxford, (2011) 1–8,
[80] Y. Saito, H. Utsunomiya, N. Tsuji, T. Sakai, Novel ultra-high straining process for bulk materials—development of the accumulative roll-bonding (ARB) process, *Acta Mater.*, 47 (2) (1999) 579–583.
[81] M. R. Toroghinejad, R. Jamaati, J. Dutkiewicz, J. A. Szpunar, Investigation of nanostructured aluminum/copper composite produced by accumulative roll bonding and folding process, *Mater. Des.*, 51 (2013) 274–279.
[82] M. I. Latypov, I. V. Alexandrov, Y. E. Beygelzimer, S. Lee, H. S. Kim, Finite element analysis of plastic deformation in twist extrusion, *Comp. Mater. Sci.*, 60 (2012) 194–200.
[83] H. Zendehdel, A. Hassani, Influence of twist extrusion process on microstructure and mechanical properties of 6063 aluminum alloy, *Mater. Des.*, 37 (2012) 13–18.
[84] Y. Beygelzimer, D. Prilepo, R. Kulagin, V. Grishaev, A. Abramova, V. Varyukhin, M. Kulakov, Planar twist extrusion versus TWIST extrusion, *J. Mater. Process. Technol.*, 211 (2011) 522–529.
[85] Y. Beygelzimer, V. Varyukhin, D. Orlov, S. Synkov, A. Spuskanyuk, Y. Pashinska, Severe plastic deformation by twist extrusion, in: M. J. Zehetbauer, R. Z. Valiev (Eds.), *Nanomaterials by Severe Plastic Deformation*, Wiley Verlag, Berlin, (2004) 511–516.
[86] V. V. Stolyarov, Y. E. Beigel'zimer, D. V. Orlov, R. Z. Valiev, Refinement of microstructure and mechanical properties of titanium processed by twist extrusion and subsequent rolling, *Phys. Met. Metallogr.*, 9 (2005) 204–211.
[87] M. I. Latypov, E. Y. Yoon, D. J. Lee, R. Kulagin, Y. Beygelzimer, M. S. Salehi, H. S. Kim, Microstructure and mechanical properties of copper processed by twist extrusion with a reduced twist-line slope, *Metall. Mater. Trans. A*, 45 (2014) 2232–2241.
[88] D. L. Schodek, P. Ferreira, M. F. Ashby, *Nanomaterials, Nanotechnologies and Design: An Introduction for Engineers and Architects*, Elsevier Science, Dordrecht, (2009).
[89] G. Kickelbick, Introduction to nanocomposites, in: M. Guglielmi, G. Kickelbick, A. Martucci (Eds.), *Sol-Gel Nanocomposites*, Springer, Berlin, (2014).
[90] S. Yu, C. J. Sun, G. M. Chow, Chemical synthesis of nanostructured particles and films, in: C. C. Koch (Ed.), *Nanostructured Materials: Processing, Properties and Applications*, (2nd edn), William Andrew Publishing, Norwich, (2007) 3–46.
[91] T. Sugimoto, X. P. Zhou, A. Muramatsu, Synthesis of uniform anatase TiO_2 nanoparticles by gel-sol method 3. formation process and size control, *J. Colloid Interf. Sci.*, 259 (2003) 43–52.
[92] T. Sugimoto, X. P. Zhou, A. Muramatsu, Synthesis of uniform anatase TiO_2 nanoparticles by gel-sol method 4. shape control, *J. Colloid Interf. Sci.*, 259 (2003) 53–61.
[93] U. Schubert, F. Schwertfeger, C. Gorsmann, Sol-gel materials with controlled nanoheterogeneity, in: G. M. Chow, K. E. Gonsalves (Eds.), *Nanotechnology: Molecularly Designed Materials*, American Chemical Society, Washington, DC, (1996) 366–381,

[94] S. Sakka, *Handbook of Sol-Gel Science and Technology: Applications of Sol-Gel Technology*, Springer, Berlin, (2005).

[95] J. D. Wright, N. A. J. M. Sommerdijk, *Sol-gel Materials: Chemistry and Applications*, CRC Press, Boca Raton, FL, (2000).

[96] F. Aldinger, V. A. Weberruss, *Advanced Ceramics and Future Materials: An Introduction to Structures, Properties, Technologies, Methods*, John Wiley & Sons, Hoboken, NJ, (2010).

[97] R. Liu, Z. M. Xie, Q. F. Fang, T. Zhang, X. P. Wang, T. Hao, C. S. Liu, Y. Dai, Nanostructured yttria dispersion-strengthened tungsten synthesized by sol-gel method, *J. Alloys Compd.*, 657 (2016) 73–80.

[98] P. K. Ting, Z. Hussain, K. Y. Cheong, Synthesis and characterization of silica–titania nanocomposite via a combination of sol–gel and mechanochemical process, *J. Alloys Compd.*, 466 (2008) 304–307.

[99] C. Suryanarayana, N. Al-Aqeeli, Mechanically alloyed nanocomposites, *Prog. Mater. Sci.*, 58(4) (2013) 383–502.

[100] K. L. Choy, C. Xu, Introduction: Chemical vapor deposition: Fundamentals and process principles, in: K. L. Choy (Ed.), *Chemical Vapour Deposition (CVD): Advances, Technology and Applications*, CRC Press, Boca Raton, FL, (2019) 1–18.

[101] L. E. Murr, Synthesis and processing of nanomaterials, in: L. E. Murr (Ed.), *Handbook of Materials Structures, Properties, Processing and Performance*. Springer, Cham, (2015) 747–766.

[102] K. L. Choy, Chemical vapour deposition of coatings, *Prog. Mater. Sci.*, 48 (2003) 57–170.

[103] D. M. Mattox, *Handbook of Physical Vapor Deposition (PVD) Processing*. Elsevier Science, Dordrecht, (2014).

[104] G. Faraji, H. S. Kim, H. T. Kashi, Introduction, in: G. Faraji, H. S. Kim, H. T. Kashi (Eds.), *Severe Plastic Deformation: Methods, Processing and Properties*, Elsevier, Oxford, (2018) 1–17.

[105] D. M. Mattox, Physical vapor deposition (PVD) processes, *Metal Finishing*, 99(Suppl. 1) (2001) 409–423.

[106] I. V. Shishkovsky, P. N. Lebedev, Chemical and physical vapour deposition methods for nanocoatings, in: A. S. H. Makhlouf, I. Tiginyanu (Eds.), *Nanocoatings and Ultra-Thin Films: Technologies and Applications*, Elsevier Science, New Delhi, (2011).

[107] A. Xia, A. Togni, S. Hirn, G. Bolelli, L. Lusvarghi, R. Franz, Angular-dependent deposition of MoNbTaVW HEA thin films by three different physical vapor deposition methods, *Surf. Coat. Technol.*, 385 (2020) 125356.

[108] H. Zeng, X-W. Du, S. C. Singh, S. A. Kulinich, S. Yang, J. He, W. Cai, Nanomaterials via laser ablation/irradiation in liquid: A review, *Adv. Funct. Mater.*, 22 (2012) 1333–1353.

[109] R. J. Varghese, E. H M. Sakho, S. Parani, S. Thomas, O. S. Oluwafemi, J. Wu, Introduction to nanomaterials: Synthesis and applications, in: S. Thomas, E. H. M. Sakho, N. Kalarikkal, S. O. Oluwafemi, J. Wu (Eds.), *Nanomaterials for Solar Cell Applications*, Elsevier, Oxford, (2019) 75–95.

[110] G. M. Mikhailov, I. V. Malikov, A. V. Chernykh, V. T. Petrashov, The effect of growth temperature on electrical conductivity and on the structure of thin refractory metal films, grown by laser ablation deposition, *Thin Solid Films*, 293(1–2) (1997) 315–319.

[111] K. Sakiyama, K. Koga, T. Seto, M. Hirasawa, T. Orii, Formation of size-selected Ni/NiO core–shell particles by pulsed laser ablation, *J. Phys. Chem. B*, 108 (2004) 523–529.

[112] Y. Kawakami, T. Seto, E. Ozawa, Characteristics of ultrafine tungsten particles produced by Nd:YAG laser irradiation. *Appl. Phys. A*, 69 (1999) S249–S252.

[113] M. Vaidya, G. M. Muralikrishna, B. S. Murty, High-entropy alloys by mechanical alloying: A review, *J. Mater. Res.*, 34(5) (2019) 664–686.

4 Synthesis of Oxide Dispersion Strengthened Refractory Alloys

4.1 EFFECTS OF OXIDE DISPERSION ON THE PROCESSING OF REFRACTORY ALLOYS

As discussed in Chapter 3, several methods can be used to process refractory alloys. The influence of oxide dispersion on the properties of refractory metals and alloys is also discussed. The processing method, such as mechanical alloying, with severe plastic deformation involving additional energy input by the addition of nano-oxides improves the process efficiency. Processing refractory metals or alloys with oxide dispersion in the nano-scale reduces the processing time with respect to final particle size requirement, by contributing additional energy for particle size refinement. Significant deformation of refractory alloys contributed by nano-oxides increases the lattice strain and the dislocation densities to improve the alloying tendency. During the synthesis of some of the refractory metals with a high melting point (W, Ta, Re, Os) and the oxides with a comparatively low melting point (Y_2O_3, Al_2O_3, La_2O_3, CeO_2) using conventional melting casting, the major problem is the segregation which significantly increases the anisotropy. The literature reports that multiple melting stages are incorporated to minimize the effect of variation in the melting temperature in the case of multicomponent synthesis, which further increases the cost [1]. As a solid-state synthesis method, mechanical alloying is proficient in synthesizing refractory alloys with a wide variation in the melting point with microstructural uniformity. Several studies on alloys such as W-Y_2O_3 [2], W-Co-Y_2O_3 [3], W-Ti-Y_2O_3 [4], W-Ni-Mn-Y_2O_3 [5], Mo-Re-La_2O_3 [6] have been carried out. The rapid solidification processing (RSP) has effectively synthesized stable and metastable alloys, though mechanical alloying is more efficient than RSP with respect to increasing the solid solubility limit [7]. The microstructural evolution may vary with the mechanical alloying of ductile/brittle (Mo/oxides) and brittle/brittle (W/oxides). Disparity in the size of the oxides and the presence of oxide particles at the interface during processing limit the final properties of the alloy [8]. Additionally, high internal energy during mechanical alloying leads to an increase in the grain size compared to the chemical mode of processing [8]. Liu et al. have suggested that the sol-gel mode of processing results in dispersion of Y_2O_3 inside the W matrix and the agglomeration tendency is reduced [9]. Liu et al. have adopted liquid-liquid processing of Mo-La_2O_3 alloys to disperse nano-sized La_2O_3 inside the Mo matrix and improvement in strength and ductility has been achieved [10]. The fabrication of complex components from powders synthesized by severe plastic deformation (SPD) is also quite challenging

[11]. The nanostructured powders tend to agglomerate and induce significant friction during compaction, which reduces the percentage relative density after compaction [11]. Dong et al. have produced a W–Y_2O_3 alloy using the hydrothermal method followed by sintering at 1600°C [12]. The developed microstructure contains flakes covered by W, distributing refined oxide particles inside the W matrix and at the boundary region [12]. Bimodal particle size distribution is an effective strategy to achieve appreciable green density (density of compact) with controlled compaction pressure and the densification during sintering can be further improved compared to monomodal size distribution. In the next section on phase evolution, the microstructure during synthesis of powder-based and bulk oxide dispersion strengthened (ODS) refractory alloys before consolidation or sintering will be discussed.

4.2 PHASE EVOLUTION IN OXIDE DISPERSION STRENGTHENED ALLOYS

During synthesis of oxide dispersion strengthened refractory alloys, the phases can be identified by X-ray diffraction (XRD), including elements and compounds. During XRD study the step size, scan rate and range of diffraction angle are important to investigate the phase formation. The lattice parameter, d spacing, structure of the phases can also be studied from the XRD pattern. The sharpness of the peak specifies the crystalline structure, whereas a diffused peak indicates the presence of the amorphous phase. In SPD and mechanical alloying (MA), the extent of deformation will regulate the final particle size. A reduction in particle size and an increase in the lattice strain is related to broadening and a decrease in the intensity of the XRD peak respectively. For amorphous alloys, instead of a sharp XRD peak (such as for crystalline material), the peak will be diffused. The formation of a solid solution can be investigated from the expansion and contraction of the lattice parameter and shifting of the XRD peak of the solvent lattice. Wang et al. have reported that the lattice parameter of W increases from 3.15293 Å (in milled pure W) to 3.15568 Å (in milled W-1.0%Y_2O_3 alloy) due to the encapsulation of Y in the W lattice [13]. Figure 4.1 shows the presence of α-W and no presence of Y_2O_3 are evident in freeze-dried W-Y_2O_3 powders with and without polymer dispersants [14].

The literature shows that during mechanical alloying of W and Y_2O_3 powder, the W peak shifts to a higher angle of diffraction, which indicates the formation of a solid solution of W [2]. The paper also reports the reduction in the lattice parameter of the solid solution of W due to Fe contamination contributed by steel grinding balls during mechanical alloying [2]. It is also reported in the paper that the peak of nano Y_2O_3 is present at 0 h of milling, which vanishes at 2 h of milling (Figure 4.2) and the crystallite size and maximum lattice strain after 50 h of milling of W-5%Y_2O_3 and W-10%Y_2O_3 powder are very close [2].

The ultrasonic spray pyrolysis-based production of W/Y_2O_3 powder involving hydrogen reduction shows the presence of α-W, β-W along with WO_3, $WO_{2.9}$, WO_2 phases at 400–800°C reduction temperature as studied by XRD [15]. No papers indicate the formation of any intermetallic between W and Y_2O_3 [2, 14, 15]. Wang et al. have synthesized W-2%ZrO_2 alloy powder by a hydrothermal technique, including hydrogen reduction, and reported the presence of W and low intensity ZrO_2 peak

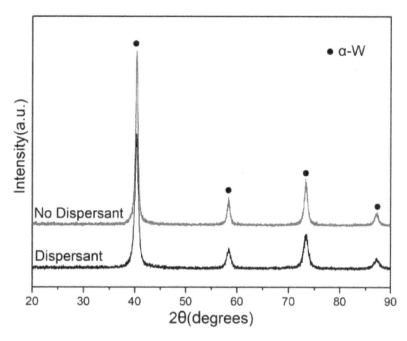

FIGURE 4.1 XRD pattern of the W-Y_2O_3 composite powders produced by freeze-drying (non-dispersion of polymer) and optimized freeze-drying (added polymer dispersants). (Reprinted from [14]. Copyright (2021), with permission from Elsevier.)

[16]. Other reports indicate that the addition of 5 wt.% Y_2O_3 in W-4 wt.% SiC does not result in the detection of the Y_2O_3 phase after 24 h of milling due to lower concentration and a broadening effect [17]. The literature reports that no traces of WC contamination from the milling media are evident with 1wt.% Y_2O_3 addition in W-4 wt.% SiC, however, a WC peak is detected in W-4 wt.% SiC-5 wt.% Y_2O_3 as a result of increased wear from the hard Y_2O_3 particles [17]. The phase volume fraction can be evaluated from the XRD pattern as [18]:

$$V_i = \frac{A_i}{\sum_{i=1}^{n} A_i} \% \quad (4.1)$$

where A_i is the diffraction peak area, V_i is the phase volume fraction.

Jakubowicz et al. have studied the variation of the lattice parameter of the Ta matrix in mechanically alloyed Ta-Y_2O_3 and Ta-ZrO_2 [19]. The paper presents that the addition of 20 wt.% of Y_2O_3 in Ta increases the lattice parameter of the Ta matrix compared to 10 wt.% Y_2O_3 addition, and the addition of 20 wt.% of ZrO_2 marginally reduces the lattice parameter of the Ta matrix against 10 wt.% of ZrO_2 addition [19]. The formation of a solid solution, amorphous phase or intermetallic in high entropy alloys depends on the enthalpy of mixing (ΔH_{mix}), the mixing entropy (ΔS_{mix}) and the difference between the atomic size (δ). A solid solution is favorable if the ΔH_{mix} is

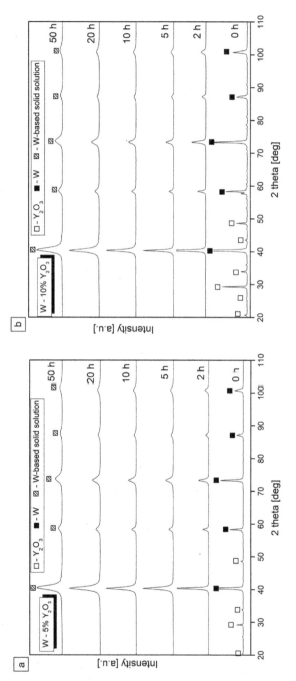

FIGURE 4.2 XRD pattern of (a) W-5%Y_2O_3 and (b) W-10%Y_2O_3 powder blend at different stages of milling. (Reprinted from [2]. Copyright (2021), with permission from Elsevier.)

Oxide Dispersion in Refractory Alloys

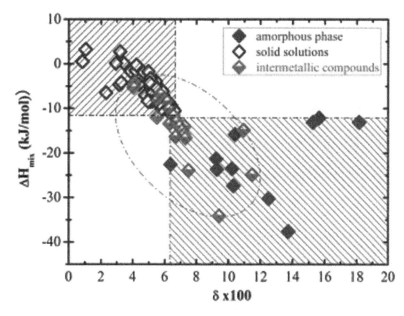

FIGURE 4.3 Graph of ΔH_{mix} versus δ illustrating the region of solid solution, intermetallic and amorphous phase formation. (Reprinted from [20]. Copyright (2013), with permission from Elsevier.)

marginally positive or negative and δ is almost less than 0.066, as illustrated in the literature (Figure 4.3) for high entropy alloys [20]. The amorphous phase develops in case the ΔH_{mix} exhibits a largely negative value, δ is greater than 0.064 and large ΔS_{mix} [21]. The intermetallic phase forms, as presented by Guo et al., are in the interim region between a solid solution and the amorphous phase [20].

4.3 MICROSTRUCTURAL CHARACTERIZATION

Microstructural development during the synthesis of the ODS refractory alloy has enormous importance with respect to its compact properties (green density) and later consolidated physical (density) and mechanical properties. The microstructure is generally studied using scanning electron microscopy (SEM) or transmission electron microscopy (TEM). Using TEM and an investigation provide an important insight into the location of the oxide dispersoids, the dislocation-particle interaction, and the extent of the matrix/oxide coherency. Selected area diffraction (SAD) and elemental mapping present the crystallinity and compositional analysis. Enhancement of the powder particle size of W-5%Y_2O_3-5%Ti after 30 h of ball milling against the initial powder particle size of W is reported by Wang et al. [13]. The paper also presents that the final structure of the ball-milled powder consists of a sphere and a sheet [13]. Synthesis of the precursor powder of W-Y_2O_3 through freeze-drying shows that an increase in cooling rate reduces the size of the flaky particle structure (less than 5 μm) and even the chemical composition (Figure 4.4) [14] and the addition of a polymer

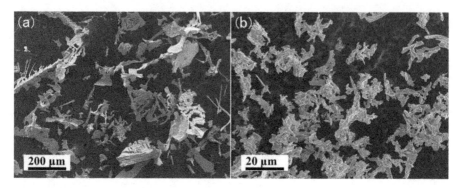

FIGURE 4.4 SEM image of precursors of W-Y_2O_3 powder synthesized through freeze-drying, (a) freezer-based cooling (low rate); (b) liquid nitrogen-based rapid cooling. (Reprinted from [14]. Copyright (2021), with permission from Elsevier.)

dispersant during freeze-drying results in a significant refinement of the grain size (mean grain size: 14 nm) [14].

Dispersion of oxide particles at the grain boundary of the matrix phase restricts the growth of the matrix grain; however, an increase in the oxides' size during subsequent heat treatment reduces the pinning pressure and the extent of the grain size refinement is less. A TEM image and energy dispersive spectroscopy (EDS) mapping of ball-milled W-5%Y_2O_3-5%Ti evidence the presence of Y, O in the nanostructured particle due to entrapment during the mechanical alloying process and, most importantly, the solute dispersion is uniform [13]. Intragranular and intergranular oxide dispersion during ODS refractory alloy synthesis are fruitful in achieving bimodal grain size distribution (due to the disparity in the grain growth) and the optimum combination of strength and ductility of the alloy can be attained.

Doping using the liquid-liquid method during synthesis of W-ZrO_2(Y), including the final reduction at 900°C, provides nano-sized and micron-sized powder particles that facilitate sintering [21, 22]. Xiao et al. have reported that dopant in pure W causes expansion of the particle size distribution compared to pure W, and homogenization of doped W alloys is presented in the paper [21]. Luo et al. have indicated that mechanical alloying of Mo and Tm_2O_3 leads to the fracturing of the brittle Tm_2O_3 particles initially against the Mo particles (3 h of milling) and an obvious reduction in Mo particle size occurs at 10 h of milling [23]. Dissociation of Tm and O followed by diffusion in Mo particles occurs at higher milling time (at 96 h of milling) [23]. An increase in milling time from 3 h to 96 h also changes the SAD pattern from a ring structure to a diffused pattern owing to the amorphous phase formation [23]. The dispersion, size and homogeneity of oxides depend on the synthesis technique. The nanoparticle spray doping method is much more effective with respect to the even distribution of La_2O_3 particles (decrease in segregation) in the grain matrix and the boundary of Mo, compared to the solid-liquid doping method [24]. The powder morphology also depends on the synthesis method. Hydrothermal synthesis of W-0.75 wt%Zr(Y)O_2 powders exhibits regular particle morphology whereas hydrothermal and mechanical alloying show a flake shape morphology as displayed in Figure 4.5

FIGURE 4.5 SEM micrographs of W-0.75 wt%Zr(Y)O_2 powders synthesized by (a) hydrothermal technique; (b) hydrothermal and mechanical alloying. (Reprinted from [25]. Copyright (2020), with permission from Elsevier.)

[25]. The flake shape morphology has considerable interparticle friction and reduces the packing effective during conventional pressureless sintering.

Sol-gel synthesis of W-1%La_2O_3 also shows crystallization of the W particles and amorphization of the oxide particles are detected from the SAD pattern [26]. Particle agglomeration in sol-gel synthesized W-1wt% La_2O_3 and W-1wt% Y_2O_3 alloy powders compared to ball-milled powders also reduces the powder compaction efficiency and green density (density of the compact before sintering) [26, 27].

It is of pivotal importance to homogenize and inhibit the agglomeration of oxide particles in the refractory matrix, to disperse oxide inside the grains and boundary to improve the effectiveness of dispersion strengthening and the enhancement of the strain hardening effect.

4.4 EFFECT OF PROCESS PARAMETERS ON MICROSTRUCTURAL DEVELOPMENT

Chen and Huang have studied the effect of grinding balls (stainless steel and tungsten carbide) on the microstructure of Y_2O_3 dispersed W-Ni-Fe alloys [28]. The particle size refinement is more significant with tungsten carbide grinding balls, due to higher density and increased collision energy transmission to the powders for particle size reduction, which also increases the lattice strain [28]. The mechanical alloying process is also influenced by the addition of a process control agent (PCA), however, the effectiveness is evident mainly with the presence of liquid PCA which inhibits the cold welding of powders [29]. Qiao et al. have illustrated that the particle morphology of high entropy TiZrNbTa refractory alloys is identical, even in the application of various PCA (solid stearic acid, liquid ethanol, liquid n-heptane) but the homogenization is enriched by the addition of liquid PCA [29]. High energy ball milling time decides the extent of the crystallinity of the alloy. Higher milling time distorts the atomic periodicity and the amorphization of the metal or alloy. Mechanical alloying of W-Y_2O_3-Ti for 30 h exhibits the formation of nanocrystalline phases which are detected from the ring-shaped SAD pattern in comparison to W where a single crystal-based particle is evident [13]. Lin et al. have studied the diffraction pattern of irradiated W-Lu_2O_3 composites by He^+ and present a diffused SAD pattern, owing to amorphization contributed by the irradiation-based atomic disruption in the lattice [30]. Different researchers have investigated the microstructure of ODS refractory alloys synthesized by mechanical alloying at different ball-to-powder weight ratios (BPR). Low BPR results in higher yield, but the particle fragmentation is less, provided the other milling parameters (speed, milling atmosphere, grinding balls, duration, wet/dry milling) are identical to the high BPR condition. During the ball milling, several factors impact the energy/unit mass in one collision (E_c) which also impacts the final particle morphology and size [31].

$$E_C = \frac{\left[7.66 \times 0^{-2} \times R_D^{\frac{1}{2}} \times \rho^{0.6} \times E^{0.4} \right] \times d_b \times \omega_D^{1/2}}{\sigma} \quad (4.2)$$

where R_D is the supportive disk radius, ρ is the grinding ball's density, E is the grinding ball's elastic modulus, d_b is the grinding ball's diameter, ω is the velocity (angular) of the disk, σ is the grinding ball casing powder's surface density.

It is clear from Equation (4.2) that an increase in ω value causes more collision energy transmission to powders and higher ρ and E value as in WC compared to stainless steel grinding balls and results in more powder particle refinement. It is also reported that the absorption of electrical powder during collision increases with an increase in the ball mill's rotational speed [31]. However, increased collision energy at high speed rotation also enhances the chances of particle agglomeration [32], which influences the particle rearrangement and density of the powder compact. Li et al. have studied the effect of grinding gas pressure during jet milling of W powder [33]. The paper reports that an increase in gas pressure from 0.55 MPa to 0.70 MPa reduces the particle coalescence and increases the particle fragmentation, sphericity [33]. Wu et al. have varied the milling speed (300, 500, 700 rpm) during ball milling of W and propose that a higher milling speed at 700 rpm reduces the particle size, however, enhancement of the milling time from 30 h to 60 h also increases the cold welding of the particles [34]. Refractory-based ODS alloys have been synthesized with different BPR during ball milling. Y_2O_3, ZrO_2, or TiO_2 dispersed W-Ni alloys have been fabricated with 10:1 BPR and low rotational speed of 80 rpm [35]. W-ZrO_2 alloy has been fabricated with 8:1 BPR and comparatively higher milling speed of 200 rpm [36]. A BPR of 4:3 and a milling speed of 400 rpm have been used for synthesis of W-Ti-La_2O_3 [37] and under these conditions the crystallite size decreases from 29 nm (at 20 h of milling) to 26.8 nm (at 30 h of milling), but surplus O content increases with the increase in the milling time [37]. The cold welding and contamination from grinding media can be controlled by optimizing the grinding speed, the BPR, and the grinding time. Hydrogen reduction at 850°C gives rise to coarse, nearly spherical and polygon-shaped W against finer W particles at 800°C reduction temperature [9]. The W particle size obtained by hydrogen reduction at 780°C (2 h) in W-1% La_2O_3 and

W-1% Y_2O_3 is 10–50 nm [26], which is comparatively more refined than the particle size of W in W-1%Y_2O_3 alloy (100–400 nm) (H_2 reduction temperature at 800°C) [9]. Chemical vapor deposited W for plasma-facing materials requires high thickness, fine grains and improved mechanical properties which can be achieved by oxide dispersion [38]. Major controlling factors for the CVD process are the deposition temperature and pressure. During irradiation in a nuclear reactor, the ODS W coating processed by CVD with a high grain boundary area accumulates vacancies and voids [38]. The impact of revolutions on the Nb microstructure during high pressure torsion (HPT) is presented by Popov and Popova and the study illustrates that the distribution of grains is unvaried, irrespective of revolutions [39]. It is described that the average grain size is 115 nm and 120 nm for 5, 10 revolutions, respectively and 5 revolutions results in high angle boundary-based crystallites [39]. Lukac et al. have also reported that an increase in revolutions from 0 to ½ in HPT of HfNbTaTiZr high entropy alloy produces ultrafine grains with coarse grains and deformation twins [40]. The diffracting domains (coherent) decrease till the revolutions number is less than 1 and a further increase in revolutions results in constancy [40]. The von Mises strain equivalent (e) is related to the number of revolutions (N), the deformed sample thickness (h) and the distance from the sample center to the strain measuring position (r) as [41]:

$$e = \frac{2\pi Nr}{h\sqrt{3}} \quad (4.3)$$

Equation (4.3) shows that the strain increases with an increase in the number of revolutions, keeping all the other parameters constant. Cizek et al. have also studied the effect of the number of revolutions on the microstructure of HfNbTaTiZr high entropy alloys at different strain levels [42]. Elevated dislocation density is reported in the grains at equivalent strain e =1, and a further increase in strain to e = 7, 25 results in a reduction in the grain size and dislocations positioned among themselves [42]. The grain size reduces to almost 80 nm when the equivalent strain e = 50, however, there is a restriction in the decrease in grain size, as e= 400 doesn't improve the grain refinement [42]. The strain rate ($\varepsilon°$) also decides the deformation mode and the refinement of the grain size during SPD and is associated with the Zener-Hollomon parameter (Z), Q (activation energy with respect to deformation), gas constant (R), and temperature (T) as [43]:

$$Z = \varepsilon° \exp\left(\frac{-Q}{RT}\right) \quad (4.4)$$

The decrease in grain size with increasing strain rate and reduced temperature is contributed by twinning which is the prevailing mechanism for low and intermediate stacking fault energy materials, while the slip is the principal deformation mechanism for high stacking fault energy materials [43]. The combination of high and low angle grain boundary and expanded angle of misorientation produces the features at elevated plastic strains [43]. HPT processed W (5 turns, HPT temperature: 500°C, pressure: 4 GPa, initial thickness: 1.0 mm, initial diameter: 10 mm) and Ta (5 turns,

HPT temperature: room temperature, pressure: ~ 5 GPa, initial thickness: 1.2 mm, initial diameter: 10 mm) show elongated W grains without any texture, whereas Ta shows equiaxed grains both exhibiting high angle grain boundaries [44, 45]. In the accumulative roll bonding (ARB) process, the rolling temperature, the rolling friction, the number of cycles in rolling, and draft in each cycle influence the microstructure and final properties. A high rolling temperature can be effective in enhancing the strength of the bond, however, recrystallization and recovery at high temperature counteract the improvement of the bond strength [46]. In ODS alloys, dispersed oxides at the grain boundary delay the recrystallization behavior and can improve the strength at high temperature applications. Diffusion rolling-based Mo-Cu composites have been developed and the microstructure shows appreciable bonding from the incidence of the interface region without any evidence of flaws, micron-sized voids or cracks [47]. The interfacial shear tensile strength of Mo/Cu developed by diffusion rolling is comparatively higher than diffusion bonding [47]. Nb-1 wt.% Zr alloy is fabricated by the ARB process and the effect of the cycle in ARB on microstructure is reported [48]. The grain size of alloy reduces from 30 μm at 0 cycle (ARB0) to 800 nm at 5 cycles (ARB5) and the microstructural changes are displayed as an electron backscatter diffraction (EBSD) study shows the major presence of high angle grain boundary (fraction: 0.9) after 5 cycles [48].

4.5 INFERENCES

The chapter presents the phase formation and microstructure of refractory grade powders produced by several methods, such as mechanical alloying, sol-gel, CVD (including thin film) and bulk samples prepared by SPD. Controlling the powders' size, shape, and distribution is of enormous importance to achieve the required final properties. Processes including compaction of nanopowders before sintering require stringent control of the powder flow properties with minimized interparticle friction during die-filling. Brittle nanopowders are more challenging to compact compared to ductile powders. Additionally, powders with high aspect ratio (flake morphology) need to be suppressed during compaction. However, the simultaneous existence of nano- and micro-sized particles provides superior packing and enhancement in compact density. In different SPD techniques (HPT, ARB), controlling the number of revolutions, the process temperature, and a reduction in each pass (percentage deformation) contribute to an impression on the microstructure and the mechanical properties and need to be regulated, based on the final applications.

REFERENCES

[1] P. Singh, A. Sharma, A. V. Smirnov, M. S. Diallo, P. K. Ray, G. Balasubramanian, D. D. Johnson. Design of high-strength refractory complex solid-solution alloys. *NPJ Comput. Mater.*, 4 (2018) 16.

[2] A. A. Dudka, D. Oleszak, R. Zielinski, T. Kulik, W-Y_2O_3 composites obtained by mechanical alloying and sintering, *Adv. Powder Technol.*, 32 (2) (2021) 390–397.

[3] C. L. C. Sutrisna, Study of W-Co ODS coating on stainless steels by mechanical alloying, *Surf. Coat. Technol.*, 350 (2018) 954–961.

[4] C. L. C. Suprianto, Investigation of W-Ti ODS coating on SUS304 steel fabricated by mechanical alloying technique, *Surf. Coat. Technol.*, 350 (2018) 1105–1111.
[5] Y. Pan, L. Ding, H. Li, D. Xiang, Effects of Y_2O_3 on the microstructure and mechanical properties of spark plasma sintered fine-grained W-Ni-Mn alloy, *J. Rare Earths*, 35 (11) (2017) 1149–1155.
[6] A. J. Mueller, R. Bianco, R.W. Buckman, Evaluation of oxide dispersion strengthened (ODS) molybdenum and molybdenum–rhenium alloys, *Int. J. Refract. Met. Hard Mater.*, 18 (4–5) (2000) 205–211.
[7] B. S. Murty, S. Ranganathan, Novel materials synthesis by mechanical alloying/milling, *Int. Mater. Rev.*, 43 (3) (1998) 101–141.
[8] Z. Dong, W. Hu, Z. Ma, C. Li, Y. Liu, The synthesis of composite powder precursors via chemical processes for the sintering of oxide dispersion-strengthened alloys, *Mater. Chem. Front.*, 3 (2019) 1952–1972.
[9] R. Liu, Z. M. Xie, Q. F. Fang, T. Zhang, X. P. Wang, T. Hao, C. S. Liu, Y. Dai, Nanostructured yttria dispersion-strengthened tungsten synthesized by sol-gel method, *J. Alloys Compd.*, 657 (2016) 73–80.
[10] G. Liu, G. J. Zhang, F. Jiang, X. D. Ding, Y. J. Sun, J. Sun, E. Ma, Nanostructured high-strength molybdenum alloys with unprecedented tensile ductility, *Nat. Mater.*, 12 (2013) 344–350.
[11] C. Ren, Z. Z. Fang, H. Zhang, M. Koopman, The study on low temperature sintering of nano-tungsten powders, *Int. J. Refract. Met. Hard Mater.*, 61 (2016) 273–278.
[12] Z. Dong, Z. Ma, J. Dong, C. Li, L. Yu, C. Liu, Y. Liu, The simultaneous improvements of strength and ductility in W–Y_2O_3 alloy obtained via an alkaline hydrothermal method and subsequent low temperature sintering, *Mater. Sci. Eng. A*, 784 (2020) 139329.
[13] R. Wang, et al., Fabrication and characterization of nanocrystalline ODS-W via a dissolution-precipitation process, *Int. J. Refract. Met. Hard Mater.*, 80 (2019) 104–113.
[14] W. Hu, Z. Dong, H. Wang, T. Ahamad, Z. Ma, Microstructure refinement and mechanical properties improvement in the W-Y_2O_3 alloys via optimized freeze-drying, *Int. J. Refract. Met. Hard Mater.*, 95 (2021) 105453.
[15] W. J. Choi, J. H. Kim, H. Lee, C. W. Park, Y. I. Lee, J. Byun, Hydrogen reduction behavior of W/Y_2O_3 powder synthesized by ultrasonic spray pyrolysis, *Int. J. Refract. Met. Hard Mater.*, 95 (2021) 105450.
[16] C. Wang, L. Zhang, S. Wei, K. Pan, X. Wu, Q. Li, Effect of ZrO_2 content on microstructure and mechanical properties of W alloys fabricated by spark plasma sintering, *Int. J. Refract. Met. Hard Mater.*, 79 (2019) 79–89.
[17] S. Coskun, M. L. Ovecoglu, Effects of Y_2O_3 additions on mechanically alloyed and sintered W-4wt.% SiC composites, *Int. J. Refract. Met. Hard Mater.*, 29 (6) (2011) 651–655.
[18] H. Dong, J. Gong, W. Li, Effect of nano-CeO_2 addition on the consolidation of W-5Ni-3Cu alloy by a two-step sintering process, *Results in Physics*, 15 (2019) 102657.
[19] J. Jakubowicz, M. Sopata, G. Adamek, P. Siwak, T. Kachlicki, Formation and properties of the Ta-Y_2O_3, Ta-ZrO_2, and Ta-TaC nanocomposites, *Adv. Mater. Sci. Eng.*, (2018), 2085368, 1–12.
[20] S. Guo, Q. Hu, C. Ng, C.T. Liu, More than entropy in high-entropy alloys: Forming solid solutions or amorphous phase, *Intermetallics*, 41 (2013) 96–103.
[21] F. Xiao, L. Xu, Y. Zhou, K. Pan, J. Li, W. Liu, S. Wei, A hybrid microstructure design strategy achieving W-ZrO_2(Y) alloy with high compressive strength and critical failure strain, *J. Alloys Compd.*, 708 (2017) 202–212.

[22] C. Han, H. Na, Y. Kim, H. Choi, In-situ synthesis of tungsten nanoparticle attached spherical tungsten micro-powder by inductively coupled thermal plasma process, *Int. J. Refract. Met. Hard Mater.*, 53 (2015) 7–12.

[23] Y. Luo, G. Ran, N. Chen, C. Wang, Microstructure and morphology of Mo-based Tm_2O_3 composites synthesized by ball milling and sintering, *Adv. Pow. Technol.*, 28 (2) (2017) 658–664.

[24] P. Feng, J. Fu, H. Zhao, X. Dang, Q. L. Yang, S. Xi, Effects of doping technologies on mechanical properties and microstructures of RE_xO_y doped molybdenum alloys, *Met. Powder Rep.*, 71 (6) (2016) 437–440.

[25] F. Xiao, Q. Miao, S. Wei, T. Barriere, G. Cheng, S. Zuo, L. Xu, Uniform nanosized oxide particles dispersion strengthened tungsten alloy fabricated involving hydrothermal method and hot isostatic pressing, *J. Alloys Compd.*, 824 (2020) 153894.

[26] R. Liu, X. P. Wang, T. Hao, C. S. Liu, Q. F. Fang, Characterization of ODS-tungsten microwave-sintered from sol–gel prepared nano-powders, *J. Nucl. Mater.*, 450 (2014) 69–74.

[27] R. Liu, Y. Zhou, T. Hao, T. Zhang, X. P. Wang, C. S. Liu, Q. F. Fang, Microwave synthesis and properties of fine-grained oxides dispersion strengthened tungsten, *J. Nucl. Mater.*, 424 (1–3) (2012) 171–175.

[28] C. L. Chen, C. L. Huang, Milling media and alloying effects on synthesis and characteristics of mechanically alloyed ODS heavy tungsten alloys, *Int. J. Refract. Met. Hard Mater.*, 44 (2014) 19–26.

[29] Y. Qiao, Y. Tang, S. Li, Y. Ye, X. Liu, L. Zhu, S. Bai, Preparation of TiZrNbTa refractory high-entropy alloy powder by mechanical alloying with liquid process control agents, *Intermetallics*, 126 (2020) 106900.

[30] J. S. Lin, L. M. Luo, Q. Xu, X. Zan, X. Y. Zhu, Y. C. Wu, Microstructure and deuterium retention after ion irradiation of $W-Lu_2O_3$ composites, *J. Nucl. Mater.*, 490 (2017) 272–278.

[31] M. Magini, A. Iasonna, F. Padella, Ball milling: An experimental support to the energy transfer evaluated by the collision model, *Scr. Mater.*, 34 (1) (1996) 13–19.

[32] R. Shashanka, D. Chaira, Optimization of milling parameters for the synthesis of nanostructured duplex and ferritic stainless steel powders by high energy planetary milling, *Powder Technol.*, 278 (2015) 35–45.

[33] R. Li, M. Qin, C. Liu, Z. Chen, X. Wang, X. Qu, Particle size distribution control and related properties improvements of tungsten powders by fluidized bed jet milling, *Adv. Powder Technol.*, 28 (2017) 1603–1610.

[34] Z. Wu, Y. Liang, E. Fu, J. Du, P. Wang, Y. Fan, Y. Zhao, Effect of ball milling parameters on the refinement of tungsten powder, *Metals*, 8 (4) (2018) 281.

[35] L. Xu, F. Xiao, S. Wei, Y. Zhou, K. Pan, X. Li, J. Li, W. Liu, Development of tungsten heavy alloy reinforced by cubic zirconia through liquid-liquid doping and mechanical alloying methods, *Int. J. Refract. Met. Hard Mater.*, 78 (2019) 1–8.

[36] F. Xiao, Q. Miao, S. Wei, Z. Li, T. Sun, L. Xu, Microstructure and mechanical properties of $W-ZrO_2$ alloys by different preparation techniques, *J. Alloys Compd.*, 774 (2019) 210–221.

[37] B. Savoini, J. Martinez, A. Muñoz, M.A. Monge, R. Pareja, Microstructure and temperature dependence of the microhardness of $W-4V-1La_2O_3$ and $W-4Ti-1La_2O_3$, *J. Nucl. Mater.*, 442 (2013) S229–S232.

[38] Z. Chen, et al., Recent research and development of thick CVD tungsten coatings for fusion application, *Tungsten*, 2 (2020) 83–93.

[39] V. V. Popov, E. N. Popova, Behavior of Nb and CuNb composites under severe plastic deformation and annealing, *Mater. Trans.*, 60 (7) (2019) 1209–1220.

[40] F. Lukac, et al., Defects in high entropy alloy HfNbTaTiZr prepared by high pressure torsion, *Acta Physica Polonica A*, 134 (3) (2018) 891–894.

[41] A. P. Zhilyaev, T. G. Langdon, Using high-pressure torsion for metal processing: Fundamentals and applications, *Prog. Mater. Sci.*, 53 (6) (2008) 893–979.

[42] J. Cizek, et al., Strength enhancement of high entropy alloy HfNbTaTiZr by severe plastic deformation, *J. Alloys Compd.*, 768 (2018) 924–937.

[43] E. Bagherpour, N. Pardis, M. Reihanian, R. Ebrahim, An overview on severe plastic deformation: Research status, techniques classification, microstructure evolution, and applications, *Int. J. Adv. Manuf. Technol.*, 100 (2019) 1647–1694.

[44] Q. Wei, H. T. Zhang, B. E. Schuster, K. T. Ramesh, R. Z. Valiev, L. J. Kecskes, R. J. Dowding, L. Magness, K. Cho, Microstructure and mechanical properties of superstrong nanocrystalline tungsten processed by high-pressure torsion, *Acta Mater.*, 54 (2006) 4079–4089.

[45] Q. Wei, Z. L. Pan, X. L. Wu, B. E. Schuster, L. J. Kecskes, R. Z. Valiev, Microstructure and mechanical properties at different length scales and strain rates of nanocrystalline tantalum produced by high-pressure torsion, *Acta Mater.*, 59 (2011) 2423–2436.

[46] S. M. Ghalehbandi, M. Malaki, M. Gupta, Accumulative roll bonding: A review, *Appl. Sci.*, 9 (17) (2019) 3627.

[47] Y. Yang, G. Lin, X. Wang, D. Chen, A. Sun, D. Wang, Fabrication of Mo-Cu composites by a diffusion-rolling procedure, *Int. J. Refract. Met. Hard Mater.*, 43 (2014) 121–124.

[48] B. L. R. Espinoza, F. A. G. Pastor, B. M. Poveda, A.R. Quesada, P. L. Crespo, High-strength low-modulus biocompatible Nb-1Zr alloy processed by accumulative roll bonding, *Mater. Sci. Eng. A*, 797 (2020) 140226.

5 Consolidation Methods

5.1 CONVENTIONAL PRESSURELESS SINTERING

This chapter introduces different consolidation techniques of refractory alloys and their respective phase and microstructure development. This first section discusses the conventional pressureless sintering involving powder compaction and sintering. The term pressureless is applied as sintering comprises no external pressure application, however, process-generated internal pressure can remain. The final sintered compact properties depend on the particle size, the compaction efficiency, the densification during compaction and sintering, the compaction load, the sintering temperature, the heating rate, and the soaking time. However, in conventional sintering, the heating rate is quite low (5°C/min) and the grain coarsening is quite high. The coarsening tendency can be restricted by dispersing oxides at the boundary. The sintering occurs in an inert atmosphere (flowing H_2 or argon) to protect the base metal/alloy from oxidation. The flowability of fine particles is comparatively less than that of the coarser particles and therefore exhibits lesser compressibility (ability of powder densification by pressure application during compaction). In pressureless sintering a lower heating rate results in shrinkage which also distorts the sample shape. Qin et al. have stated that the grain size of pure W and W-1wt% La_2O_3 alloy sintered at 1200°C for 2 h in a flowing H_2 atmosphere is 0.57 μm and 0.28 μm, respectively with faceted grain morphology [1]. With an increase in sintering temperature to 1350°C the corresponding grain sizes are 3.95 μm for pure W and 0.43 μm for W-1wt% La_2O_3 alloy [1]. The results indicate prevention of grain coarsening by dispersed La_2O_3 particles [1]. The lowering of the sintering temperature for ODS refractory alloys depends on the extent of nanostructuring before densification, which is also important to control the grain growth. Lee et al. have adopted sintering of ODS W alloys in two phases (solid-state and liquid-phase sintering) [2]. The solid-state sintering was carried out at 1300°C –1450°C (1 h holding) in H_2 and the liquid-phase sintering at 1465°C –1485°C for 30 min in an argon atmosphere [2]. The grain size of the liquid-phase sintered W at 0 h of sintering time and 1485°C temperature is 1.8 μm which significantly enhances to 19.8 μm with an increase in the holding time to 1 h at a similar temperature [2]. However, the W/W contiguity reduces with an increase in holding time and a similar trend is observed with the increase in the liquid-phase sintering temperature from 1465°C to 1485°C (30 min holding time) [2]. Reduction in contiguity of W/W is effective in order to improve the toughness of material. The contiguity of the W phase can be evaluated as [3]:

$$C_W = \frac{2N_{WW}}{N_{WM} + 2N_{WW}} \quad (5.1)$$

where N_{WW} is the W/W contacts number/unit length of a certain intercept, N_{WM} is the W/matrix contacts number/unit length of a certain intercept.

Reports also suggest that the augmented addition of Ni in W-Cr-Ni reduces the contiguity by enhancing the mass transport facilitated by the Ni-based liquid phase [3]. Pressureless sintering can also be applied to fabricate crucibles, turbine blades, and rocket guidance fins [4]. Production of complex designs through this method is challenging owing to the shrinkage effect [4]. The linear shrinkage $\left(\dfrac{\Delta L}{L_0}\right)$ is presented as [5]: where

$$\frac{\Delta L}{L_0} = \left(\frac{K\gamma D^* a^3}{kTd^n}\right) m \quad (5.2)$$

where

$\Delta L/L_0$: linear shrinkage, a: diffusion vacancy's atomic volume, γ: surface energy, K: shape factor, D^*: coefficient of self-diffusion, t: time, k: Boltzmann constant, d: diameter of particle, and T: temperature. Equation 2 shows that finer particle size before sintering enriches the rate of shrinkage. Higher the available time for holding during conventional pressureless sintering higher will be the shrinkage rate. Still, the finer particles with higher grain boundary area will boost the mass transfer during sintering. Li et al. have used two-step pressureless sintering technique to produce fine grain, microstructural homogeneity, and constricted grain size distribution for W [6]. The first sintering step includes 1300–1450 °C with 1 h of holding followed by the second step sintering at 1200–1350 °C with 10 h of holding [6]. Some grain growth has been reported in the paper and more impact is provided on improving the green density (density of compact before sintering) by controlling the particle size and shape [6]. The temperature effect on grain size during sintering is given by scaling law [7,8]:

$$n \ln\left(\frac{d_1}{d_2}\right) = \frac{Q}{R}\left(\frac{1}{T_2} - \frac{1}{T_1}\right) \quad (5.3)$$

where d_1 is the particle size at temperature T_1, d_2 is the particle size at temperature T_2, Q is the activation energy, R is the gas constant. The grain size variation with temperature sintered W-Ti alloy in Ar and H_2 environment is studied in the literature [9]. When the sintering temperature is less in the H_2 environment, the vapor transport is effective and with increasing temperature beyond 1200°C, the volume diffusion is established as the sintering mechanism (Figure 5.1) [9]. The paper also reports that volume and grain boundary diffusion are significant during sintering in the Ar environment [9].

If the difference in melting point between the refractory metals and the added alloying constituents is significant, then pressureless sintering is challenging. Therefore, nanoparticles can be added with micron-sized refractory powder for milling and sintering [10]. Qingxiang et al. have fabricated W-20wt.%Ti by milling

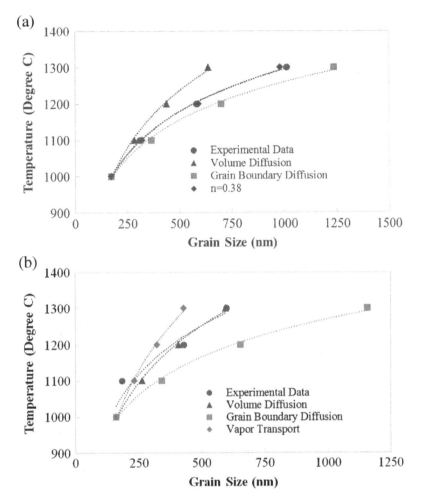

FIGURE 5.1 Sintering temperature and grain size plot for nano W-1 wt.%Ti illustrated from scaling law in sintered (a) Ar and (b) H_2 environment. (Reprinted from [9]. Copyright (2017), with permission from Elsevier.)

of nanostructured TiH_2 powders and micron-based W powders followed by uniaxial compaction and sintering in a vacuum environment at 1200°C [10]. The paper also reports that adding nanostructured TiH_2 facilitates the diffusion of Ti in W, a reduction of the sintering temperature and microstructural homogeneity [10]. The sintering atmosphere is decided by the composition, such as in W-V, W-Nb, W-Ti and W-Ta, the hydrogen embrittlement is evident, but the problem can be counteracted by the formation of inert (Argon) or a vacuum environment at the final sintering period [11, 12]. High temperature pressureless sintering (1800°C, for 2 h) in a hydrogen atmosphere increases the grain coarsening (grain size: 24.8 μm) compared to hot pressing (1800°C, for 2 h) (grain size: 8.8 μm) and spark plasma sintering (1600°C for 3 min)

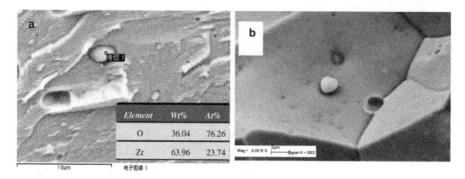

FIGURE 5.2 (a) EDS study, (b) SEM study of Mo-0.1Zr alloy. (Reprinted from [13]. Copyright (2009), with permission from Elsevier.)

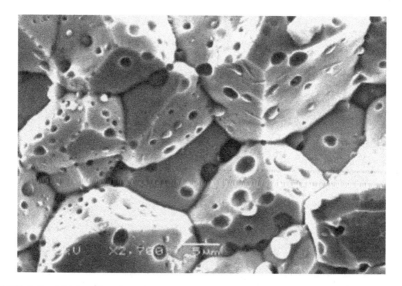

FIGURE 5.3 SEM study of Mo-0.8Ti alloy with 0.8 wt% TiH_2 addition. (Reprinted from [13]. Copyright (2009), with permission from Elsevier.)

(grain size: 2.3 μm) carried out in a vacuum [12]. High temperature sintering of Mo-Ti and Mo-Zr starting with 1920°C–1980°C (3 h, H_2 environment) has been carried out by Fan et al. [13]. It is reported that oxides of Zr are present at the grain boundary and matrix (Figure 5.2) [13], which negatively impacts the tensile strength [14] and oxides of Mo, Ti is located at the Mo grain boundary (Figure 5.3 and Figure 5.4) [13].

5.2 SPARK PLASMA SINTERING

Spark plasma sintering (SPS) involves sintering at an elevated heating rate with nominal holding time in a vacuum environment with the assistance of external pressure.

Consolidation Methods

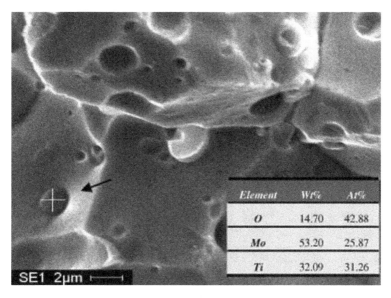

FIGURE 5.4 EDS study of particles located at grain boundary of Mo-0.8Ti alloy with 0.8 wt% TiH$_2$ addition. (Reprinted from [13]. Copyright (2009), with permission from Elsevier.)

Due to less sintering time, the grain growth is considerably suppressed. The powders are placed inside a graphite die or conducting die and pulsed direct current (DC) (low voltage) is passed (for conductive powders) through the powders [15], as displayed in Figure 5.5 [16]. Non-conductive powders, if placed in the conductive mold, also result in faster particle bonding and densification [17]. For sintering of small samples, SPS is an effective method with enriched densification, but with larger samples (size > 50 mm), temperature fluctuation results in uneven densification and microstructure [16], which can be counteracted by the adoption of double heating techniques [18, 19]. The heating rate during SPS can be as high as 1000°C/min, however, it depends on the type of sample and die dimension, though increasing the rate of heating creates an uneven temperature between the powder and the die [20]. Observation of the die temperature is possible through an IR pyrometer and thermocouple [20]. Step-wise increase in heating rate is a solution to prevent the thermal shock and cracking of sintered products. During sintering, a graphite paper spacer is also used to facilitate the current flow to the die-punch contact [21, 22]. One of the major challenges in sintering nanostructured materials is preserving the nanosize during sintering, which can be accomplished by SPS with minimal soaking time and sintering temperature.

SPS includes an electrical release from a pulsed DC and spark formation at the junction of the particles of powders, resulting in high heat generation (> 1000°C). The vaporization of impurities and particle surface melting occur from the elevated temperature. The particle neck development takes place through captivating of melting parts [20]. The applied pressure makes a significant contribution in stimulating the densification and establishing particle contacts by grain boundary sliding and the plastic deformation mechanism [23]. Angerer et al. have reported that increasing

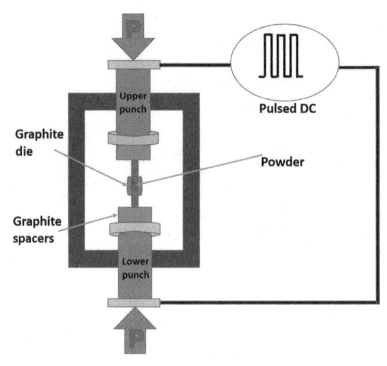

FIGURE 5.5 Schematic of SPS setup. (Reprinted from [16]. Copyright (2019), with permission from Elsevier.)

the SPS temperature from 1500°C to 1700°C with 1 min of holding time at sintering temperature and 100°C/min heating rate for Ta powder marginally increases the crystallite size from 20 nm to 22 nm [24]. Further increase in SPS temperature to 1900°C with identical holding time and heating rate significantly increases the crystallite size to 113 nm and additionally, the impurity content (Ta_2C) also increases with the increase in temperature [24]. A similar trend is also reported in the literature throughout SPS of Ru at 1200°C, 1400°C, 1600°C (heating rate: 100°C/min, holding time: 1 min) [25]. Wiedemann et al. have reported sintering of Mo by SPS at various heating rates (130–360 K/min), temperature (850°C–2000°C) and pressure (29–67 MPa) [26]. The morphology of the particles is maintained till 1100°C, however, at higher temperatures (>1400°C), polyhedral morphology of particles is reported [26]. The paper also presents the influence of temperature at a constant pressure of 67 MPa on grain size in terms of mean chord length (Figure 5.6) and indicates that enhancement of grain size occurs with an increase in SPS temperature beyond 1330°C [26].

In SPS, several diffusion techniques are active, such as evaporation-condensation, surface and melt diffusion, which is relatively rapid which reduces the sintering temperature by 200–300°C compared to conventional sintering techniques [27]. SPS techniques involving a low voltage indicate a release of the spark, elevated temperature plasma leading to a joule heating effect and the enrichment of atomic diffusion and densification [27]. During nanostructuring, the grain boundary possesses

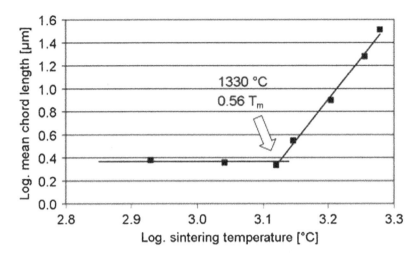

FIGURE 5.6 Plot of Log (grain growth) and Log (sintering temperature) molybdenum needs (pressure: 67 MPa). (Reprinted from [26]. Copyright (2010), with permission from Elsevier.)

enhanced concentration of vacancies which are transported in the grains and facilitate atomic transport during the SPS of W [28]. It is also reported that the apparent activation energy during the SPS of W at 1500°C is 41±1 kJ/mol [28]. The addition of a ductile constituent such as 3wt.% Re in W improves the plastic flow of W and further decreases the activation energy to 38 ± 1 kJ mol^{-1} [28]. The shrinkage characteristics of pressure-assisted sintering of W-Ti powders mean that the rate of shrinkage increases till 900°C followed by a decrease till 1500°C and further attainment of constancy [29]. The increase in shrinkage rate is attributed to neck generation and growth owing to a rise in temperature and current [29]. A drop in the shrinkage rate due to restrictions in the rearrangement of particles and the enhancement of the atomic diffusion length contributes to constancy in the shrinkage rate with an increase in temperature beyond 1500°C [29].

In the pressure- and temperature-assisted SPS process, the rate of creep $(\dot{\varepsilon})$ can be expressed as [30]:

$$\dot{\varepsilon} = \frac{A\Phi\mu_{eff}b}{kT}\left(\frac{b}{G}\right)^p \left(\frac{\sigma_{eff}}{\mu_{eff}}\right)^n \quad (5.4)$$

where A is the constant, Φ is the diffusion coefficient, μ_{eff} is the instantaneous shear modulus of the sintered compact, b is Burgers vector, G is the grain size, k is the Boltzmann constant, T is the absolute temperature, σ_{eff} is the effective instantaneous stress functional to the sintered compact, p is the exponent of grain size, n is the effective stress exponent.

The value of n decides the mode of densification at different temperature regimes. Huang et al. have indicated that during the SPS of W-Ti alloy the applied pressure at the starting temperature of 900°C, 1000°C, 1100°C and n < 1 results in reorganization of the particles [29]. In the case of n = 1–2 during the advanced stage of 900°C, 1000°C, 1100°C and at a starting temperature of 1200°C, 1300°C temperature, the grain boundary diffusion is operative, whereas at the advanced stage of 1200°C, 1300°C and n = 3.77/4.14, the dislocation climb is operative [29]. Liquid-phase sintering during SPS adds challenges for effective ejection owing to sticking in the die. However, using graphite cloth in the die also leads to graphite deposited on the sample surface which needs additional grinding for further microstructural study. Low temperature sintering, refined grain size, nearly retaining the sample dimension, and enormous densification are the prime value additions in the SPS process.

5.3 MICROWAVE SINTERING

Microwave sintering is an advanced material consolidation method and has several benefits, such as (1) reduced processing time; (2) improvement in diffusion; (3) fine grain size; (4) lower energy consumption; and (5) it is an environmentally sound process compared to conventional sintering [31]. The heating in microwave sintering is associated with microwave field absorption and transforming electromagnetic energy into thermal energy for volumetric heating of the component [31]. Figure 5.7 illustrates the difference between conventional and microwave sintering. In microwave sintering, the heat is generated inside the materials and flows outside and the heating during microwave sintering depends on the microwave absorption capacity [31]. The frequency in microwave sintering is normally 2.45 GHz and the depth of penetration is significant compared to conventional sintering [32]. Microwave

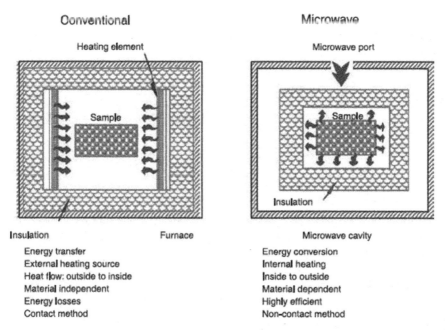

FIGURE 5.7 Heating methods of conventional and microwave sintering. (Reprinted from [31], Copyright (2010), with permission from Elsevier.)

sintering can be carried out in different environments, such as in 90%N_2 + 10%H_2 [33, 34], 5%H_2 and 95%Ar [35]. The inert atmosphere prevents any unwanted oxygen diffusion from developing oxides which impair the mechanical properties. The selection of the H_2 or N_2 atmosphere also depends on the affinity of the metal/alloy to generate brittle hydrides or nitrides formation. The temperature during microwave sintering can be examined by an infrared pyrometer [36].

Rodiger et al. have reported that the time requirement for heating and soaking in microwave sintering is less by a factor of 3 compared to conventional sintering and, without the use of any pressure, hard metals, cermets (binder content: 25 wt.% or 4 wt.%) can be sintered by microwave sintering at 50–100 K lower temperature than that required for conventional sintering [32]. Microwave sintering (whole heating rate: 20°C/min) of 92.5W–6.4Ni–1.1Fe alloy has been reported in the literature with a decrease in time by 75% against conventional sintering (heating rate: 5°C/min) [37]. The microwave sintered alloy also exhibits no evidence of NiW, Fe_7W_6 intermetallic phases, refined W grain size (9.4 ± 0.5 μm) but higher contiguity (0.42 ± 0.09) than the presence of the intermetallics, reduced contiguity (0.32 ± 0.10), and coarser W grains (17.3 ± 0.8 μm) in conventionally sintered alloy [37]. The heating rate in microwave sintering also influences the grain size and morphology. It is reported that the W grain size in W-Ni-Fe at 10°C/min heating rate is 13.6 μm and reduces to 9.6 μm at 112°C/min [33]. Figure 5.8 shows that the shape of W grains is impacted by

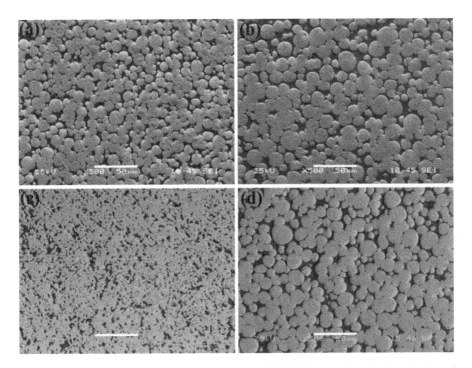

FIGURE 5.8 SEM images of microwave sintered W-Ni-Fe samples with (a) 0 min and (b) 10 min holding time and heating rate 29°C/min, (c) 0 min and (d) 10 min holding time and heating rate 65°C/min at 1480°C. (Reprinted from [33], Copyright (2009), with permission from Elsevier.)

the holding time and the heating rate [33]. Increasing the heating rate from 29°C/min to 65°C/min without a holding time results in a departure from the spherical shape of W at 1480°C owing to minimizing the matrix dissolution of W at 65°C/min [33]. However, maintaining the holding time of 10 min at both heating rates shows a spherical shape [33].

Microwave sintering of W-Y_2O_3 and W-HfO_2 has been carried out in a hydrogen atmosphere at 1400°C with 20 min of holding and reduces the grain size to 0.5 μm [38]. Liu et al. have consolidated W-Ni-Fe alloy by microwave sintering with a heating rate of 30°C/min at 1250°C, 1300°C, 1350°C, 1400°C, 1450°C, 1500°C and 5 min of holding time [36]. Significant grain growth occurs beyond 1350°C and the shape of the grains also depends on the sintering temperature and sintering mechanism [36]. Mainly polygonal W grains have been reported at 1400°C and, with further increase in sintering temperature, this facilitates liquid-phase sintering and the grain shape changes to a spherical shape at 1500°C [36]. An increased dissolution and diffusion mechanism during microwave sintering also hinders the brittle intermetallic phase formation [36].

5.4 HOT ISOSTATIC PRESSING

During uniaxial compaction, the punch moves in one direction inside the die and therefore removing the pressure/density gradient throughout the sample cross-section is challenging. In isostatic pressing the pressure is homogeneously transferred to all directions of the sample and the isostatic pressing, carried out at a high temperature involving gas-based pressure transmission for significant densification, is denoted as hot isostatic pressing (hipping/HIP) [39]. It is reported that during hipping, gas atoms (velocity: 900 m/s) collide with the sample surface (10^{30} collisions/square meter/s), which develops the isostatic pressure [39]. Therefore, the microporosity, macroporosity and deviation in average values of properties are minimized [39]. The pressure in hipping can be 70–300 MPa, as well as higher than 1 GPa is also employed, but the applied temperature is reduced by 25–50% compared to isothermal sintering methods [40]. Different cycles can be adopted in the hipping process, such as standard HIP and sinter followed by the HIP method [40]. In standard HIP, the powder is subjected to preforming by cold uniaxial or isostatic pressing, injection molding followed by transfer to an evacuated can, subsequent HIP process, and finally the sample is removed from the container [40]. In the sinter and HIP process, a preform is developed by cold uniaxial or isostatic pressing, injection molding, hot forming and high temperature sintering followed by HIP to fabricate complex geometries [40]. The sinter HIP process can reduce the process and equipment cost [41]. Apart from that, several other techniques, such as HIP quenching involving elevated cooling (regulated cooling rate: 500°C/min) can result in the reduction of the processing time and distortion [42], and liquid-HIP/quick-HIP with oil/grease as the visco-plastic medium [43]. HIP of W-Ta has been carried out at 1528 K (1255°C) for 2 h in an Ar atmosphere and 200 MPa pressure [44]. Before HIP, the composites were subjected to deoxidation at 1373 K (1100°C), for 3 h, followed by canning and degassing at 723 K (450°C) for 14 h with Ar [44]. The paper reports that W matrix size in the hipped W-Ta composites is < 200 nm and the solid solution of W-Ta is hindered by the formation of Ta_2O_5 [44]. A minimum W grain size of 2.6 μm is reported in hipped W-ZrO_2 alloy compared to hydrogen furnace sintering, hot vertical sintering and the hot swaging process [45]. Martinez et al. have reported microcrack development in V regions with surface relief facilitated by thermal stress development during the cooling cycle of hipped W-4%V-1%La_2O_3 with grain size < almost 0.5 μm [46]. HIP also facilitates diffusion bonding between W/W and TZM/TZM (Molybdenum-titanium-zirconium) with interfacial homogeneity, less imperfection and bonding strength of 215 MPa, 550 MPa has been reported [47]. The investigation also shows that the diffusion bonding of Ta2.5W/TZM is a 100% efficient joint, with trivial thermal resistance at the interface [47]. The overall benefits of HIP include a homogeneous structure, isotropic properties, reduced cost of machining, improved toughness, and the development of composite products [48], but the cost of the equipment can be a challenge which is counteracted by its superior properties [49].

5.5 HYDROSTATIC EXTRUSION

In hydrostatic extrusion (HE), the fluid with elevated pressure transfers the hydrostatic pressure for the extrusion of a billet in non-contact mode (as in traditional extrusion) and the friction between the billet and die is reduced by the fluid and ductility is enhanced [50]. The major components of the hydrostatic extrusion are presented in Figure 5.9 [50].

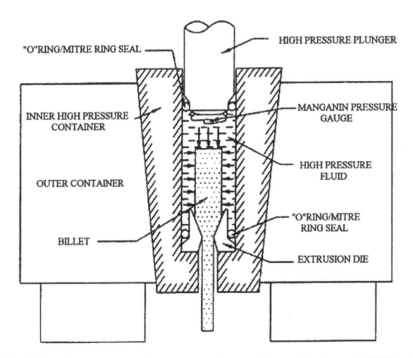

FIGURE 5.9 Hydrostatic extrusion apparatus. (Reprinted from [50]. Copyright (2000), with permission from Elsevier.)

The high pressure fluid can generate pressure of 10 000 kg/cm^2 (about 10 000 atm pressure) in the container and therefore the container design can be based on two layers [50]. The literature reports the use of cost-effective SKD61 tool steel as the container material to hold the pressure [50]. Container sealing is provided by the plunger and the movement of the plunger causes the fluid pressure development and the billet is subjected to move through the die [50]. Controlling the material flow is required without friction to inhibit deflection and distortion of samples [51]. The major benefits of hydrostatic extrusion include fabrication of compound products, attainment of an elevated extrusion ratio, and dimensional precision, however, the complex equipment, reduced efficiency of materials and restricted fabrication of shapes are a few challenges [51]. The die angle in HE influences the frictional force as an increase in the die angle decreases the friction force while a higher extrusion ratio (starting area of cross-section/final area of cross-section) enhances the extrusion pressure related to the elevated friction coefficient and yield strength [52]. Li et al. have fabricated W-40 wt.%Cu composites by milling and liquid-phase sintering at 1100°C–1200°C followed by hot hydrostatic extrusion at 800°C, 850°C, 900°C, 950°C in a die which is preheated at 250°C [53]. During extrusion, the temperature is of key importance, as the increase in temperature leads to reduced hydrostatic pressure as the extrusion pressure drops [53]. Microstructural uniformity with 6–10 μm grain size of W is reported in the literature [53]. Zhang et al. have hydrostatically extruded 93W-4.9Ni-2.1Fe tungsten heavy alloy with 28%, 59% and 85% deformation and note that the W grain size transforms to an ellipsoid structure with the increase in the percentage deformation, and the presence of a fibrous structure is reported after 85% of deformation with respect to the sintered grain size, as displayed in Figure 5.10 (a–d) and the W grain's axial contiguity also decreases with the enhancement of the percentage deformation [54]. The W/W grain boundaries are the weak links and a decrease in axial contiguity improves the alloy strength [54]. The reduction of contiguity is related to the permeation of the matrix in the W/W grain boundary owing to the superior plasticity and pressure effect in hydrostatic extrusion [55, 56].

Higher strength of W and deformation ability are added advantages in hydrostatic extrusion compared to the swage method (the common deformation method). In single-pass hydrostatic extrusion, deformation of 60–80% is accomplished compared to swage where the percentage deformation of > 20% results in a rise in the rate of crashing [57].

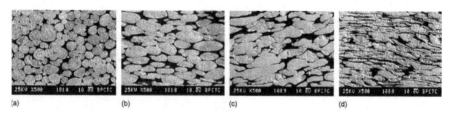

FIGURE 5.10 Structure of 93W alloy (a) in sintered condition and hydrostatic extrusion with (b) 28% deformation, (c) 59% deformation, (d) 85% deformation. (Reprinted from [54]. Copyright (2006), with permission from Elsevier.)

5.6 POWDER FORGING

Traditional forging results in a product with a high density, however, it suffers from inadequate dimensional precision and post-forging machining [58]. In powder forging, the combined application of compressive and shear deformation results in densification with near net shape added with minimal machining or even without the need of any machining, in the absence of flash generation [58]. In the process, the powder is compacted and preheated or sintered, followed by heating to a temperature for forging and then further forged in a die of the required shape [58]. The literature on manufacturing of refractory alloys through powder forging is not extensive. Few reports indicate how to fabricate a disc of milled, vacuum hot pressed and the subsequent high energy rate forging and annealing for residual stress elimination [59]. The purity of the powder is important to control the ductility and the impact strength during the forging operation [60]. During the fabrication of preform, the density and the powder particle size distribution must be adequately controlled to enhance the response to forging [60]. In powder forging, cold and warm compaction, assisted by a multidirectional press, facilitates the fabrication of complex geometries (gears, sprockets) as well as the bulk production of connecting rods and gear rings [61]. It is also reported that developed connecting rods by the powder forging route possess elevated strength, density and durability [62]. Phillips et al. have reported that the production cost of powder forged components is enhanced by 50–75% compared to warm compaction and double pressing–sintering-based production [63], which impose challenges, though the incidence of fine grain size, a steady dislocation substructure and the nominal deterioration of reinforced constituent [64, 65]. The procedure adopted in powder forging of flanged parts is presented in Figure 5.11, which illustrates how the preform undergoes compaction in the sleeve die assisted by the

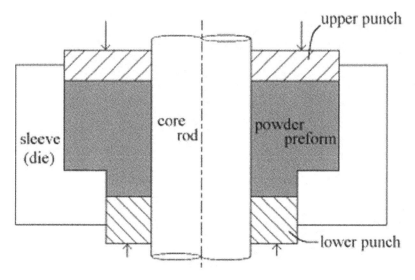

FIGURE 5.11 Representation of powder forging method for flanged parts. (Reprinted from [66]. Copyright (2001), with permission from Elsevier.)

punch set-up, and the final density (relative) will be around unity [66]. The selection of the die in the powder forging process is very important in order to control the stress development in the final product. In the case of isothermal forging, use of a costly die (Ni superalloys, Mo alloys) and lubricants can be offset by powder metallurgy-based preform development [67, 68].

TiNbTa0.5ZrAl0.5 refractory (high entropy alloy) has been formed by blending, cold isostatic pressing (200 MPa), vacuum sintering (1300°C, 16 h, vacuum level: 1×10^{-3} Pa) and cooling in a furnace followed by hot forging at 1200°C [69]. The hot-forged microstructure exhibits a BCC matrix and the HCP precipitates with the homogeneous presence of precipitates and the matrix size of 1 µm [69]. Mo-ZrO_2 powders are compacted by cold isostatic pressing (280 MPa, hold time: 5 min) and H_2-based sintering (2000°C) has been carried out to fabricate rods (diameter: 20 mm), and finally hot die forging (1000°C) resulted in a bar shape after polishing [70, 71]. The forging process leads the second-phase particles (size: ~ 200 nm) to be refined and ZrO_2 particles can be observed at the grain boundary and grain interior (Figure 5.12 (a, b)) [70]. Dislocation loop formation occurs by the resistance of the ZrO_2 particles, which improves the strengthening effect by grain boundary strengthening and intragranular strengthening mode [70].

During forging of powder preforms the selection of the forging process, the pressure and strain effects, and the initial density of the powder preform influence the deformation and densification behavior. The density of the preform is significant when a higher strain is applied and the pressure ratio to achieve a minimum level and maximum preform density to achieve 95% densification is 1·75:1 [72]. In closed die forging, the impact of the density of the preform with respect to preventing the lateral flow of the material is less compared to plane strain forging where the deformation characteristics of the porous preform and the friction effect are dominant [72]

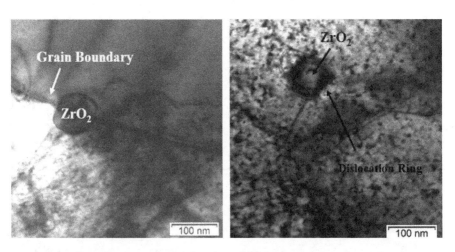

(a) Grain boundary strengthening (b) Intragranular strengthening

FIGURE 5.12 TEM image of Mo-1.5wt.%ZrO_2 alloy bar. (Reprinted from [70]. Copyright (2020), with permission from Elsevier.)

Consolidation Methods 105

Additionally, in closed die forging, the major challenge to achieve enhanced densification is contributed by the slow development of the die wall friction. During forging in an open die, the variation in preform and forged component volume will reduce, caused by the starting uniaxial deformation [72]. During forging, the appearance of the density gradient, due to non-uniform deformation/pressure owing to shape and friction effects and therefore tool design, the selection of the forging type, and the preform fabrication all need to be taken into account [72].

5.7 SINTER FORGING

In sinter forging, sintered parts are forged to enhance the densification, thus an enhanced yield and product shape can be obtained as the final product [73]. The sinter forging method is sometimes called powder forging, and the latter is more common with the forging of powder preforms. Uniaxial pressure application in sinter forging facilitates the free material flow in the lateral direction as no die wall is involved [74]. Simple geometries (a plate shape) can be fabricated by this process [74]. The hydrostatic stress for free upsetting and closed die forging is $\sigma/3$, $\sigma-2\gamma/3$ (where σ is axial compressive stress, γ is resistance to deformation) and it is evident that due to the absence of the deformation resistance component in free upsetting/rotary forging, the pores can be collapsed more effectively and superior density can be achieved [75]. The homogeneous deformation of the forged product can be achieved by either improving the density of the preform or by rotary forging as low preform density can create a barrelling problem [75]. In conventional sinter forging, the compact is placed on a fixed section made of ceramic, and a punch (ceramic) can apply the load to the sample [76]. The separation between the punch and the bottom part with the sample is made by placing carbon plates to inhibit sticking [76]. Advanced sinter forging can be microwave sinter forging (MWSF) as represented in Figure 5.13 [76] or electro sinter forging (ESF) [77]. In MWSF, the cleaned green compact is placed between one punch (it can be made of alumina) and a steel rod (water-cooled) setup, each at the top and the bottom. The bottom setup is fixed, whereas the top setup can move to incorporate the required force in the compact [76]. It is also stated that the sample shape depends on the applied stress in MWSF. Below 17 MPa stress, the sample is cylindrical and the shape changes to barrel when the stress is below 30 MPa [76]. The paper also reports that shrinkage in a radial direction is not inhibited in conventional sinter forging which is possible in MWSF and the mean grain size achieved in MWSF (420 nm) is less than the grain size of microwave sintered sample (510 nm) [76]. In ESF, strong electric pulses and a current density of $0.1- > 1$ kA/mm^2 are applied by the plunger within a time frame of 30–100 milliseconds and both mechanical and electrical pulses are incorporated into the die [77]. The major reported benefits of ESF are: process automation, enhanced mechanical properties with reduced tolerance, it eliminates the need for a vacuum/gas environment, however. samples with less complexity, low weight (< 200 gm), and cold dies cannot be used for sintering of ceramic materials which offers a challenge in the process [77].

The fewer process steps in sinter forging, the superior isotropic properties compared to conventional forging, decreased forging pressure, lower temperature and reduction in percentage of scrap all bring down the overall processing cost, which makes sinter

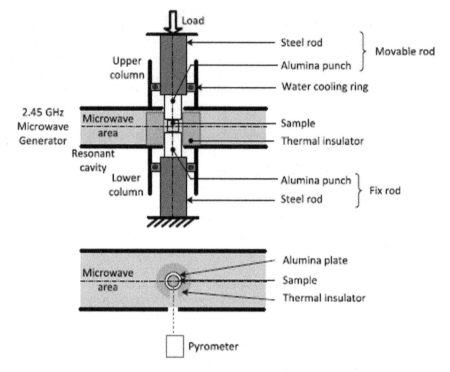

FIGURE 5.13 Upper image: Front view, Bottom image: Top view of microwave sinter forging setup. (Reprinted from [76]. Copyright (2015), with permission from Elsevier.)

forging a candidate fabrication method [78]. Amosov et al. have reported that particle size fraction and powder purity decide the forging characteristics and therefore need to be precisely regulated, and the forging reduction ratio marginally deviates as the powder purity enhances [79]. Mo billets produced by pressing and a sintering technique can easily be forged compared to ingots produced by arc casting, due to the columnar grain formation during arc casting, which imparts brittleness during forging [80]. For W, the forgeability depends on the density and temperature as a reduction in density and enhancing the forging temperature reduce the forgeability, therefore, the production of billets with a diameter > 3″ is quite challenging [80]. In this context, the powder compaction prior to forging is highly significant and can be carried out by hydrostatic or isostatic pressing instead of a mechanical/hydraulic press [80]. For adequate forgeability, the density at the middle of the billet should be at least 90% and a mean density of 93–94% of theoretical density is desired [80]. During forging, a protective environment needs to be ensured for pressed and sintered W, W alloys (W-1Th, W-15Mo), Mo, Mo alloys (Mo-30W, Mo-50W) which are very prone to oxidation [80].

5.8 SHOCK-BASED CONSOLIDATION

Consolidation of powders by shock waves involves a high velocity effect or an explosion resulting in elevated pressure (>1 GPa) and a strain rate-based (10^7 to around

Consolidation Methods

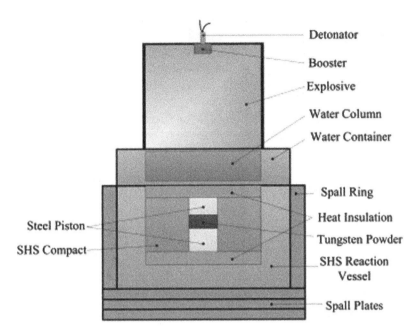

FIGURE 5.14 Representation of HSC facilitated by SHS. (Reprinted from [81]. Copyright (2014), with permission from Elsevier.)

10^8 s^{-1}) densification [81]. The phenomenon leads to enhancing the cooling rate (10^9 K/s) near the surface [82] and assists in surface melting with the cold inner surface of the particle [83]. Minimum fabrication time (within a microsecond), bulk consolidation and the prevention of recrystallization and grain growth are some of the benefits of this process [81, 84, 85]. The hot shock consolidation (HSC) is displayed in Figure 5.14 [81]. In the process, the powder is heated by the reaction through self-propagating high-temperature synthesis (SHS) prior to the shock wave consolidation. The heat insulator (ZrO$_2$ plates) is placed at the top and bottom part of the sample to minimize the heat loss, the excessive heating of the explosives and the water column [81]. Ignition of the explosive (nitro-methane) is carried out by the detonator, followed by cooling and machining of the consolidated sample [81].

W-25%Cu alloy has been fabricated by dynamic consolidation (detonation velocity: 5200 m/s) and, after consolidation, a homogeneous microstructure without any intermetallic formation and a W crystallite size of 27 nm is reported in the literature [86]. The literature evidences that both pressure and preheating temperature are important factors to achieve appreciable sintering characteristics, as a low preheating temperature results in inadequate densification, whereas elevated pressure improves bonding but increases cracking [81]. The report also suggests that pressure has a leading effect in order to enrich the sintering behavior [81]. W-7 wt.%Ni-3 wt.% Fe alloy has been consolidated by shock waves with a preheating temperature before consolidation between 300°C and 1000°C and a shock wave-based pressure of 10 GPa [87]. The W grain size is 50.7 µm (standard deviation: 10 µm) at 300°C, 29.4

μm (standard deviation: 4.2 μm) at 400°C, 58 μm (standard deviation: 11.9 μm) at 1000°C and the microstructure in the process compared to conventional consolidation is identical [87]. The rapid rate of consolidation in this process leads to low contamination [87]. Several studies of hot explosive consolidation-based fabrication of W-Ti alloys [88] and Mo-Ti alloys [89] have been reported. The W-Ti alloys exhibit a segregation effect owing to the faster cooling effect and the matrix uniformity can be achieved by the temperature of the sample (near to melting point of Ti) [88]. In case of Mo-Ti alloys, the inner region shows deviation from the initial composition (90Mo-10Ti, all in wt%), some areas of the peripheral region exhibit homogeneity with respect to base composition [89]. Compared to W, the reduced thermal diffusivity of Mo does not produce an acceptable matrix equilibrium of β-Ti/Mo [90, 89]. The reported Mo grain size is 10–30 μm with transverse orientation and the coring effect of Mo-Ti (can be reduced by heat treatment less than the monotectoid temperature) [89]. Fabrication of W-Cu alloys with microstructural homogeneity is challenging [91], owing to the elevated positive heat of mixing 35.5 kJ mol^{-1} [92], which leads to low solubility in W and Cu [91]. Reports state the fabrication of W-10 wt.%Cu alloy by mechanical alloying followed by compaction, preheating at different temperatures (between minimum: 702°C, maximum: 1028°C) and hot shock consolidation (pressure of shock: 3.1 GPa, 4.3 GPa) [93]. The paper specifies that with the increase in the preheating temperature, homogenization improves and so does the movement of the Cu along the vacant spaces of W particles, which also improves the bonding [93]. The above reports show that shock-based consolidation is an alternative method of alloy fabrication, but the particle size, preheating temperature, pressure during shock wave application all need to be controlled to achieve a crack-free, homogeneous microstructure.

5.9 HIGH PRESSURE SINTERING

High pressure sintering is immensely important in fabricating fine-grained metals/alloys/compounds. In several cases, the high pressure induces phase transformation by facilitating nucleation and decreases the rate of diffusion, therefore, the rate of grain growth is also reduced [94]. High pressure sintering is beneficial as per the following aspects [95]:

1. reduced sintering temperature and time;
2. cost-effective, specifically for high melting point refractory metals; the reduction in sintering temperature with appreciable mechanical properties is highly desired in this particular context;
3. prevention of grain coarsening and addition of dispersed oxides in refractory metals/alloys further improves grain refinement;
4. facilitates workability.

In high pressure sintering, the rate of activation energy required for plastic deformation ($\dot{\varepsilon}$) is related to the free energy change (normal to activated condition) (ΔF), volume subjected to activation (ΔV), where A is the constant, P is the pressure, R is

Consolidation Methods

the universal gas constant, T is the temperature, and the plastic deformation gradually enhances with an increase in pressure [96]:

$$\dot{\varepsilon} = A \exp[-(\Delta F - \Delta VP)/RT] \tag{5.5}$$

It is also important to note that the diffusion reduces during enrichment of pressure in the sintering [97]:

$$D = D_0 \exp[-\Delta VP/RT] \tag{5.6}$$

Elevated pressure increases the density of dislocation along with sintering at low temperature which results in higher diffusion through dislocation rather than a lattice and, therefore, low temperature-based power law creep is the operative mechanism [98]. Increased pressure enhances the particle contact and the particle neck grows, followed by an increase in the particle co-ordination number, which assists in bonding [99]. When the pressure is high (in the GPa range), the temperature is low with a high yield strength, and at a low temperature the grain growth is hindered. However, the pressure effect on densification is minimized as the temperature is increased [99]. For nanostructured materials sintered by high pressure sintering, the grain boundary diffusion is dominant, owing to the increased grain boundary area [99] and if the pressure and temperature of sintering are high, the volume diffusion and dislocation motion are triggered [100, 101, 102]. High pressure has been applied to fabricate $(W_{0.5}Al_{0.5})C_{0.5}$ with 4.5 GPa pressure, at a temperature of 1600°C (10 min) and it is reported that an increase in sintering time (5–30 min) during high pressure sintering at 1600°C has a minimum influence on the grain coarsening, and the dominant factor for grain size control is the sintering temperature [103]. The literature also describes how W can be sintered by elevated pressure (2–9 GPa) and resistance sintering (electrical current-based sintering) and such high pressure-induced sintering are denoted as ultra-high pressure sintering [104]. The powder particles undergo rearrangement, sliding, distortion and get crushed by such a high pressure effect [104]. The high pressure application also reduces the coalescence of particles facilitating plastic/superplastic deformation-based densification [105]. Kuntz et al. have developed Ta-WO_3 with the application of high pressure (300 MPa) and at a low temperature of 500°C during SPS [106]. During such high pressure SPS, the die material is graphite, the SiC plunger and supporting disc of WC are incorporated [105]. Liu et al. have studied the effect of high pressure (4.5 GPa, 5.5 GPa) on the microstructure of sintered W (at 1000°C, 1300°C, 1400°C. for 30 min) [107]. The presence of finer particles has been reported at 4.5 GPa (1000°C), though at pressure, at the temperature of 4.5 GPa, 130°C respectively, damaged grains are evident, and, at increased pressure of 5.5 GPa with 1300°C sintering temperature, the cracked grains are evident [107]. Ultra high pressure (9000 MPa) assisted resistance sintering (1300°C, power: 23 kW) of Mo exhibits a grain size of 2.12 μm and the sintering time is also relatively less (1 min) [108]. High pressure sintering facilitates the consolidation of refractory metals/alloys at much lower temperature compared to their high melting point with reduced time, but the design and selection of the material for the die, anvil, plunger and pressure level are vital to refine the grain with less distortion.

5.10 INFERENCES

Different powder consolidation techniques are described in the above sections. Application of pressure and temperature has an immense effect in controlling the fine grain size, homogeneity, and the grain shape. The pressure can either be applied prior to sintering or during sintering. Nanostructured powders are susceptible to agglomeration and, therefore, flowability should be taken into account to achieve enhanced packing efficiency. The grain growth rate can be impeded by dispersing oxides and increasing the heating rate and pressure. The high pressure during sintering can also minimize the agglomeration tendency of the powders and improve the sintering kinetics. A low sintering time is also beneficial to control the shrinkage effect in the final consolidated product, to reduce the machining and increase the yield. Apart from pressure, the sintering temperature, time, and environment are quite important to reduce the chances of oxide formation in the refractory grade materials. Production of complex shapes is also possible with HIP, the powder forging process. The effectiveness of any consolidation process depends on its overall energy consumption (in terms of sintering temperature, sintering time), the diversity in the fabrication of shapes, the die-punch set-up and the materials involved, the sintering atmosphere (inert/vacuum) as well as the powder characteristics (flow property, particle size, particle shape) which need to be modulated to improve the sintering behavior.

REFERENCES

[1] M. Qin, et al., Preparation of intragranular-oxide-strengthened ultrafine-grained tungsten via low-temperature pressureless sintering, *Mater. Sci. Eng. A*, 774 (2020) 138878.

[2] K. H. Lee, S. I. Cha, H. J. Ryu, S. H. Hong, Effect of two-stage sintering process on microstructure and mechanical properties of ODS tungsten heavy alloy, *Mater. Sci. Eng. A*, 458 (1–2) (2007) 323–329.

[3] V. Srikanth, G. S. Upadhyaya, Contiguity variation in tungsten spheroids of sintered heavy alloys, *Metallography*, 19 (1986) 437–445.

[4] P. C. Angelo, R. Subramanian, *Powder Metallurgy: Science, Technology and Applications*, PHI Learning Pvt. Ltd., New Delhi, (2009).

[5] D. W. Richerson, W. E. Lee, *Modern Ceramic Engineering: Properties, Processing, and Use in Design* (4th edn.), CRC Press, Boca Raton, FL, (2018).

[6] X. Li, L. Zhang, Y. Dong, R. Gao, M. Qin, X. Qu, J. Li, Pressureless two-step sintering of ultrafine-grained tungsten, *Acta Mater.*, 186 (2020) 116–123.

[7] G. L. Messing, M. Kumagai, Low-temperature sintering of alpha-alumina-seeded boehmite gels, *Am. Ceram. Soc. Bull.*, 73 (1994) 88–91.

[8] M. F. Yan, W. W. Rhodes, Low temperature sintering of TiO_2, *Mater. Sci. Eng.*, 61 (1983) 59–66.

[9] C. Ren, M. Koopman, Z. Z. Fang, H. Zhang, B. V. Devener, A study on the sintering of ultrafine grained tungsten with Ti-based additives, *Int. J. Refract. Met. Hard Mater.*, 65 (2017) 2–8.

[10] W. Qingxiang, L. Shuhua, F. Zhikang, C. Xin, Fabrication of W-20wt.%Ti alloy by pressureless sintering at low temperature, *Int. J. Refract. Met. Hard Mater.*, 28 (5) (2010) 576–579.

[11] A. Bose, R. M. German, Sintering atmosphere effects on tensile properties of heavy alloys, *Metall. Trans. A*, 19 (1988) 2467–2476.

[12] K. Arshad, J. Wang, Y. Yuan, Y. Zhang, Z. J. Zhou, G. H. Lu, Development of tungsten-based materials by different sintering techniques, *Int. J. Refract. Met. Hard Mater.*, 50 (2015) 253–257.

[13] J. Fan, M. Lu, H. Cheng, J. Tian, B. Huang, Effect of alloying elements Ti, Zr on the property and microstructure of molybdenum, *Int. J. Refract. Met. Hard Mater.*, 27 (1) (2009) 78–82.

[14] Y. Q. Li, J. Y. Liu, *Interstitial Phase in the Grain Boundary of Refractory Alloy*, (1st edn.), Metallurgy Industry Press, Beijing (1990).

[15] A. V. Ragulya, Fundamentals of spark plasma sintering, in: K. H. Jurgen Buschow, R. W. Chan, M. C. Flemings, B. Ilschner, E. J. Kramer, S. Mahajan, S. Mahajan, P. Veyssiere (Eds.), *Reference Module in Materials Science and Materials Engineering*, Elsevier, Oxford, (2016) 1–5.

[16] M. R. Mphahlele, E. A. Olevsky, P. A. Olubambi, Spark plasma sintering of near net shape titanium aluminide: A review, in: G. Cao, C. Estournes, J. Garay, R. Orru (Eds.), *Spark Plasma Sintering*, Elsevier, Oxford, (2019) 281–299.

[17] M. Tokita, Recent and future progress on advanced ceramics sintering by spark plasma sintering, *Nanotechnol. Russ.*, 10 (3–4) (2015) 261–267.

[18] M. B. Shongwe, S. Diouf, M. O. Durowoju, P. A. Olubambi, M. M. Ramakokovhu, B. A. Obadele, A comparative study of spark plasma sintering and hybrid spark plasma sintering of 93W–4.9 Ni–2.1 Fe heavy alloy, *Int. J. Refract. Met. Hard Mater.*, 55 (2016) 16–23.

[19] F. Vogeler, B. Lauwers, E. Ferraris, Analysis of wire-EDM finishing cuts on large scale ZrO_2-TiN hybrid spark plasma sintered blanks, *Procedia CIRP*, 42 (2016) 268–273.

[20] P. Cavaliere, B. Sadeghi, A. Shabani, Spark plasma sintering: process fundamentals, in: P. Cavaliere (Ed.), *Spark Plasma Sintering of Materials*, Springer, Cham, (2019) 3–20.

[21] R. Yamanoglu, N. Gulsoy, E. A. Olevsky, H. O. Gulsoy, Production of porous Ti_5Al_2.5Fe alloy via pressureless spark plasma sintering, *J. Alloys Compd.*, 680 (2016) 654–658.

[22] W. L. Bradbury, E. A. Olevsky, Production of SiC–C composites by free-pressureless spark plasma sintering (FPSPS), *Scr. Mater.*, 63 (2010) 77–80.

[23] O. Guillon, J. J. Gonzalez, B. Dargatz, T. Kessel, G. Schierning, J. Rathel, M. Herrmann, Field-assisted sintering technology/spark plasma sintering: mechanisms, materials, and technology developments, *Adv. Eng. Mater.*, 16 (2014) 830–849.

[24] P. Angerer, E. Neubauer, L. G. Yu, K. A. Khor, Texture and structure evolution of tantalum powder samples during spark-plasma-sintering (SPS) and conventional hot-pressing, *Int. J. Refract. Met. Hard Mater.*, 25 (4) (2007) 280–285.

[25] P. Angerer, J. Wosik, E. Neubauer, L.G. Yu, G. E. Nauer, K. A. Khor, Residual stress of ruthenium powder samples compacted by spark-plasma-sintering (SPS) determined by X-ray diffraction, *Int. J. Refract. Met. Hard Mater.*, 27 (1) (2009) 105–110.

[26] R. O. Wiedemann, U. Martin, H. J. Seifert, A. Muller, Densification behaviour of pure molybdenum powder by spark plasma sintering, *Int. J. Refract. Met. Hard Mater.*, 28 (4) (2010) 550–557.

[27] Z. Y. Hu, Z. H. Zhang, X. W. Cheng, F. C. Wang, Y. F. Zhang, S. L. Li, A review of multi-physical fields induced phenomena and effects in spark plasma sintering: Fundamentals and applications, *Mater. Des.*, 191 (2020) 108662.

[28] S. Pramanik, A. K. Srivastav, B. M. Jolly, N. Chawake, B. S. Murty, Effect of Re on microstructural evolution and densification kinetics during spark plasma sintering of nanocrystalline W, *Adv. Powder Technol.*, 30 (11) (2019) 2779–2786.

[29] L. Huang, J. Zhang, Y. Pan, Y. Du, Spark plasma sintering of W-10Ti high-purity sputtering target: Densification mechanism and microstructure evolution, *Int. J. Refract. Met. Hard Mater.*, 92 (2020) 105313.

[30] A. S. Helle, K. E. Easterling, M. F. Ashby, Hot-isostatic pressing diagrams: new developments, *Acta Metall.*, 33 (1985) 2163–2174.

[31] D. Agrawal, Microwave sintering of ceramics, composites and metal powders, in: Z. Z. Fang (Ed.), *Sintering of Advanced Materials*, In Woodhead Publishing Series in Metals and Surface Engineering, Cambridge, (2010) 222–248.

[32] K. Rodiger, K. Dreyer, T. Gerdes, M. W. Porada, Microwave sintering of hardmetals, *Int. J. Refract. Met. Hard Mater.*, 16 (1998) 409–416.

[33] C. Zhou, J. Yi, S. Luo, Y. Peng, L. Li, G. Chen, Effect of heating rate on the microwave sintered W-Ni-Fe heavy alloys, *J. Alloys Compd.*, 482 (1–2) (2009) L6–L8.

[34] R. Liu, T. Hao, K. Wang, T. Zhang, X. P. Wang, C. S. Liu, Q. F. Fang, Microwave sintering of W/Cu functionally graded materials, *J. Nucl. Mater.*, 431 (1–3) (2012) 196–201.

[35] K. Wang, X. P. Wang, R. Liu, T. Hao, T. Zhang, C. S. Liu, Q. F. Fang, The study on the microwave sintering of tungsten at relatively low temperature, *J. Nucl. Mater.*, 431 (1–3) (2012) 206–211.

[36] W. Liu, Y. Ma, J. Zhang, Properties and microstructural evolution of W-Ni-Fe alloy via microwave sintering, *Int. J. Refract. Met. Hard Mater.*, 35 (2012) 138–142.

[37] A. Upadhyaya, S. K. Tiwari, P. Mishra, Microwave sintering of W-Ni-Fe alloy, *Scr. Mater.*, 56 (1) (2007) 5–8.

[38] M. Jain, G. Skandan, K. Martin, D. Kapoor, K. Cho, B. Klotz, R. Dowding, D. Agrawal, J. Cheng, Microwave sintering: A new approach to fine-grain tungsten – II, *Int. J. Powder Metall.*, 42 (2) (2006) 53–57.

[39] H.V. Atkinson, S. Davies, Fundamental aspects of hot isostatic pressing: An overview. *Metall. Mater. Trans. A*, 31 (2000) 2981–3000.

[40] R. F. Davis, Hot isostatic pressing, in: R. J. Brook (Ed.), *Concise Encyclopedia of Advanced Ceramic Materials*, Pergamon, Oxford, (1991) 210–215.

[41] N. L. Loh, K. Y. Sia, An overview of hot isostatic pressing, *J. Mater. Process Technol.*, 30 (1992) 45–65.

[42] A. Traff, New developments in hot isostatic press (HIP) units, *Met. Powder Rep.*, 40 (1) (1990) 279–283.

[43] T. Matsushita, K. Notomi, N. Kawai, T. Yamasaki, Hot isostatic compaction of tool steel powders by visco-plastic pressure medium, *R&D Kobe Steel Eng. Rep.*, 40 (1) (1990) 38–41.

[44] M. Dias, F. Guerreiro, J. B. Correia, A. Galatanu, M. Rosinski, M. A. Monge, A. Munoz, E. Alves, P.A. Carvalho, Consolidation of W-Ta composites: Hot isostatic pressing and spark and pulse plasma sintering, *Fusion Eng. Des.*, 98–99 (2015) 1950–1955.

[45] F. Xiao, Q. Miao, S. Wei, Z. Li, T. Sun, L. Xu, Microstructure and mechanical properties of W-ZrO_2 alloys by different preparation techniques, *J. Alloys Compd.*, 774 (2019) 210–221.

[46] J. Martinez, B. Savoini, M. A. Monge, A. Munoz, R. Pareja, Development of oxide dispersion strengthened W alloys produced by hot isostatic pressing, *Fusion Eng. Des.*, 86 (2011) 2534–2537.

[47] J. B. Descarrega, C. Torregrosa, M. Calviani, Development and beam irradiation of Ir/W/Ta/Ta-alloys refractory metals and cladding via hot isostatic pressing at CERN for beam intercepting devices applications, in: *Proc. 14th Int. Workshop Spallation Materials Technology*, *JPS Conf. Proc.*, 28 (2020) 041002.

[48] C. Broeckmann, Hot isostatic pressing of near net shape components – process fundamentals and future challenges, *Powder Metall.*, 55 (3) (2012) 176–179.

[49] A. R. Bhatti, P. M. Farries, Preparation of long-fiber-reinforced dense glass and ceramic matrix composites, in: A. Kelly, C. Zweben (Eds.), *Comprehensive Composite Materials*, Pergamon, Oxford, (2000) 645–667.

[50] J. C. Hung, C. Hung, The design and development of a hydrostatic extrusion apparatus, *J. Mater. Process. Technol.*, 104 (3) (2000) 226–235.

[51] W. H. Sillekens, J. Bohlen, Hydrostatic extrusion of magnesium alloys, in: C. Bettles, M. Barnett (Eds.), *Advances in Wrought Magnesium Alloys*, Woodhead Publishing, Cambridge, (2012) 323–345.

[52] J. J. Lewandowski, A. Awadallah, Hydrostatic extrusion of metals and alloys, in: S. L. Semiatin (Ed.), *ASM Handbook*, vol. 14A, *Metalworking: Bulk Forming*, ASM, Materials Park, OH, (2005), 440–447.

[53] D. Li, Z. Liu, Y. Yu, E. Wang, Research on the densification of W-40 wt.%Cu by liquid sintering and hot-hydrostatic extrusion, *Int. J. Refract. Met. Hard Mater.*, 26 (4) (2008) 286–289.

[54] Z. H. Zhang, F. C. Wang, S. K. Li, L. Wang, Deformation characteristics of the 93W-4.9Ni-2.1Fe tungsten heavy alloy deformed by hydrostatic extrusion, *Mater. Sci. Eng. A*, 435–436 (2006) 632–637.

[55] J. W. Noh, E. P. Kim, H. S. Song, W. H. Baek, K. S. Churn, S. J. L. Kang, Matrix penetration of the W/W grain boundaries, *Metall. Trans. A*, 24 (11) (1993) 2411–2416.

[56] J. B. Posthill, M. C. Hogwood, D. V. Edmonds, Precipitation at tungsten/tungsten interfaces in tungsten-nickel-iron heavy alloys, *Powder Metall.*, 29 (1) (1986) 45–51.

[57] W. Huanyu, C. Hongnian, Research of the strengthening technology by hydrostatic extrusion for tungsten alloy, *Adv. Mater. Manu. Technol.*, 5 (1998) 1–5.

[58] W. J. Huppmann, M. Hirschvogel, Powder forging, *Int. Met. Rev.*, 23 (1) (1978) 209–239.

[59] Y. Tan, Y. Y. Lian, F. Feng, Z. Chen, J. B. Wang, X. Liu, W. G. Guo, L. Cheng, G. H. Lu, Surface modification and deuterium retention of high energy rate forging W-Y$_2$O$_3$ exposed to deuterium plasma, *J. Nucl. Mater.*, 509 (2018) 145–151.

[60] J. S. Hirschhorn, R. B. Bargainnier, The forging of powder metallurgy preforms, *JOM*, 22 (1970) 21–29.

[61] V. Y. Dorofeev, Y. G. Dorofeev, Powder forging: Today and tomorrow, powder metall. *Met. Ceram.*, 52 (2013) 386–392.

[62] J. M. Capus, PM Tec 2004—is North American PM back on track?, *Powder Metall.*, 43 (3) (2004) 226–228.

[63] R. R. Phillips, D. Hammond, I. L. Friedman, PM aims for direct competition with 'old-tech' industry, *Met. Powder Rep.*, 9 (2004) 26–35.

[64] R. Narayanasamy, Hot forging of sintered steel–titanium carbide composites, *Mat. Des.*, 29 (7) (2008) 1380–1400.

[65] G. A. Baglyuk, I. D. Martyukhin, T. M. Pavligo, G. G. Serdyuk, V. M. Tkach, Structural features of hot-forged carbide steel (high-speed steel–titanium carbide, *Powder Metall. Met. Ceram.*, 48 (1–2) (2009) 34–37.

[66] J. R. Cho, Y. S. Joo, H. S. Jeong, The Al-powder forging process: Its finite element analysis, *J. Mater. Process. Technol.*, 111 (1–3) (2001) 204–209.

[67] M. Chandrasekaran, Forging of metals and alloys for biomedical applications, in: M. Niinomi (Ed.), *Metals for Biomedical Devices* (2nd edn), Woodhead Publishing, Cambridge, (2019) 293–310.

[68] I. Marek, P. Novak, A. J. Mlynar, D. Vojtcha, T. F. Kubatik, J. Malek, Powder metallurgy preparation of Co-based alloys for biomedical applications, *Acta Phys. Pol. A.*, 128 (4) (2015) 597–601.

[69] Y. Cao, Y. Liu, Y. Li, B. Liu, A. Fu, Y. Nie, Precipitation behavior and mechanical properties of a hot-worked TiNbTa0.5ZrAl0.5 refractory high entropy alloy, *Int. J. Refract. Met. Hard Mater.*, 86 (2020) 105132.

[70] C. Chaopeng, Z. Xiangwei, L. Qiang, Z. Min, Z. Guangping, L. Shulong, Study on high temperature strengthening mechanism of ZrO_2/Mo alloys, *J. Alloys Compd.*, 829 (2020) 154630.

[71] C. Cui, Y. Gao, S. Wei, G. Zhang, X. Zhu, S. Guo, Preparation and properties of ZrO_2/Mo alloys, *High Temp. Mater. Process.*, 36 (2) (2017) 163–166.

[72] H. F. Fischmeister, B. Aren, K. E. Easterling, Deformation and densification of porous preforms in hot forging, *Powder Metall.*, 14 (27) (1971) 144–163.

[73] N. Chawla, J. J. Williams, R. Saha, Mechanical behavior and microstructure characterization of sinter-forged SiC particle reinforced aluminum matrix composites, *J. Light Metals*, 2 (4) (2002) 215–227.

[74] A. J. A. Winnubst, M. M. R. Boutz, Sintering and densification; New techniques: sinter forging, *Key Eng. Mater.*, 153–154 (1998) 301–323.

[75] P. M. Standring, E. Appleton, Rotary forging development in Japan II. Theoretical investigation and analysis: Powder compaction and sinter-forging, *J. Mech. Work. Technol.*, 4 (1) (1980) 7–29.

[76] V. Delobelle, J. Croquesel, D. Bouvard, J. M. Chaix, C. P. Carry, Microwave sinter forging of alumina powder, *Ceram. Int.*, 41 (6) (2015) 7910–7915.

[77] A. Fais, A faster FAST: Electro-sinter-forging, *Met. Powder Rep.*, 73 (2) (2018) 80–86.

[78] S. Kumar, A. K. Jha, Present status and future potential of sinter-forging technology, in: F. D. Marquis (Ed.), *Powder Materials: Current Research and Industrial Practices*, vol. III, John Wiley & Sons, Hoboken, NJ, (2010).

[79] V. M. Amosov, N. N. Bobkova, V. V. Dianov, The dependence of the technological properties of tantalum and niobium on the physicochemical characteristics of their starting powders. *Powder Metall. Met. Ceram.*, 4 (1965) 534–537.

[80] H. J. Henning, F. W. Boulger, *Notes on the Forging of Refractory Metals*, Vol. 143 of DMIC memorandum, Defense Metals Information Center, Battelle Memorial Institute, The University of Michigan, Ann Arbor, MI, (1961).

[81] Q. Zhou, P. Chen, Fabrication and characterization of pure tungsten using the hot-shock consolidation, *Int. J. Refract. Met. Hard Mater.*, 42 (2014) 215–220.

[82] W. H. Gourdin, Energy deposition and microstructural modification in dynamically consolidated metal powders, *J. Appl. Phys.*, 55 (1984) 172.

[83] T. G. Nieh, P. Luo, W. Nellis, D. Lesuer, D. Benson, Dynamic compaction of aluminum nanocrystals, *Acta Mater.*, 44 (1996) 3781–3788.

[84] G. E. Korth, R. L. Williamson, Dynamic consolidation of metastable nanocrystalline powders, *Metall. Mater. Trans. A*, 26 (1995) 2571–2578.

[85] K. P. Stuadhammer, K. A. Johnson, L. E. Murr, K. P. Stuadhammer, M. A. Meyers (Eds.), *Metallurgical Application of Shock-wave and High-strain-rate Phenomena*, Marcel Dekker, New York, (1986), 149.

[86] Z. Wang, X. Li, J. Zhu, F. Mo, C. Zhao, L. Wang, Dynamic consolidation of W-Cu nanocomposites from W-CuO powder mixture, *Mater. Sci. Eng. A*, 527 (21–22) (2010) 6098–6101.

[87] F. D. S. Marquis, A. Mahajan, A. G. Mamalis, Shock synthesis and densification of tungsten based heavy alloys, *J. Mater. Process. Technol.*, 161 (2005) 113–120.

[88] L. J. Kecskes, I. W. Hall, Hot explosive consolidation of W-Ti alloys, *Metall. Mater. Trans. A*, 26 (9) (1995) 2407–2414.

[89] L. J. Kecskes, Hot explosive compaction of Mo-Ti alloys, *Metall. Mater. Trans. A*, 30 (9) (1999) 2483–2489.

[90] Y. S. Touloukian, R. W. Powell, C. Y. Ho, M. C. Nicolaou (Eds.), *Thermal Diffusivity*, IFI/Plenum, New York, (1973) 113, 198.

[91] C. Hou, X. Song, F. Tang, Y. Li, L. Cao, J. Wang, Z. Nie, W-Cu composites with submicron- and nanostructures: Progress and challenges, *NPG Asia Mater.*, 11 (2019) 74.

[92] T. Raghu, R. Sundaresan, P. Ramakrishnan, T. R. R. Mohan, Synthesis of nanocrystalline copper–tungsten alloys by mechanical alloying, *Mater. Sci. Eng. A*, 304–306 (2001) 438–441.

[93] Q. Zhou, P. Chen, Characterization of fine-grained W–10wt.% Cu composite fabricated by hot-shock consolidation, *Int. J. Refract. Met. Hard Mater.*, 52 (2015) 137–142.

[94] Z. Z. Fang, H. Wang, Sintering of ultrafine and nanosized ceramic and metallic particles, in: R. Banerjee, I. Manna (Eds.), *Ceramic Nanocomposites*, Woodhead Publishing, Cambridge, (2013) 431–473.

[95] K. Liu, D. He, H. Wang, T. Lu, F. Li, X. Zhou, High-pressure sintering mechanism of yttrium aluminum garnet (Y3Al5O12) transparent nanoceramics, *Scr. Mater.*, 66 (6) (2012) 319–322.

[96] T. Yokobori, *An Interdisciplinary Approach to Fracture and Strength of Solids*, Metallurgiya, Moscow, (1971).

[97] P. G. Shewmon, *Diffusion in Solids*, McGraw-Hill Book Co., London, (1963).

[98] J. Trapp, A. Semenov, M. Nothe, T. Wallmersperger, B. Kieback, Fundamental principles of spark plasma sintering of metals: part III – densification by plasticity and creep deformation, *Powder Metall.*, 63 (5) (2020) 329–337.

[99] R. M. German, Sintering with external pressure, in: R. M. German (Ed.), *Sintering: From Empirical Observations to Scientific Principles*, Butterworth-Heinemann, Oxford, (2014) 305–354.

[100] M. F. Ashby, Background reading HIP 6.0, Engineering Department Cambridge University, Cambridge, (1990).

[101] Y. S. Kwon, K. T. Kim, Densification forming of alumina powder-effects of power law creep and friction, *J. Eng. Mater. Tech.*, 118 (1996) 471–477.

[102] W. R. Cannon, T. G. Langdon, Review creep of ceramics, Part 2: An examination of flow mechanisms, *J. Mater. Sci.*, 23 (1988) 1–20.

[103] J. Yan, X. Ma, W. Zhao, H. Tang, C. Zhu, S. Cai, High-pressure sintering study of a novel hard material $(W_{0.5}Al_{0.5})C_{0.5}$ without binder metal, *Int. J. Refract. Met. Hard Mater.*, 25 (1) (2007) 62–66.

[104] Z. Zhou, Y. Ma, J. Du, J. Linke, Fabrication and characterization of ultra-fine grained tungsten by resistance sintering under ultra-high pressure, *Mater. Sci. Eng. A*, 505 (1–2) (2009) 131–135.

[105] U. A. Tamburini, J. E. Garay, Z. A. Munir, Fast low-temperature consolidation of bulk nanometric ceramic materials, *Scr. Mater.*, 54 (2006) 823–828.

[106] J. D. Kuntz, O. G. Cervantes, A. E. Gash, Z. A. Munir, Tantalum–tungsten oxide thermite composites prepared by sol-gel synthesis and spark plasma sintering, *Combustion and Flame*, 157 (8) (2010) 1566–1571.

[107] P. Liu, et al., High-pressure preparation of bulk tungsten material with near-full densification and high fracture toughness, *Int. J. Refract. Met. Hard Mater.*, 42 (2014) 47–50.

[108] Z. Zhou, N. Deng, H. Wang, J. Du, Fabrication of fine grained molybdenum by fast resistance sintering under ultra-high pressure, *J. Alloys Compd.*, 782 (2019) 899–904.

6 Densification of Consolidated Products

6.1 MECHANISM OF DENSIFICATION

The densification of a consolidated product can occur during sintering as in the case of pressure-based sintering or a combination of densification prior to sintering (during the fabrication of green compacts) and in sintering relevant to pressureless sintering. The densification also depends on the involvement of phases, either solid (solid state sintering), or liquid (liquid phase sintering). Solid-state sintering occurs below the melting point of base metal or any alloying constituent. The reduction in the overall surface energy occurs due to grain growth which triggers the sintering process. The sintering process involves three steps: (1) initial step: the neck develops and grows (radius = 0– ~0.2 of particle radius) as represented in Figure 6.1 (a), followed by the neck coinciding; (2) intermediate step: the neck growth continues and the pore network is persistent (Figure 6.1 (b)); and (3) final step: isolated pore development which is positioned at the confluence of four grains (Figure 6.1 (c)) and the process occurs as the percentage relative sintering density is almost 92% [1]. Different modes of diffusion are operative at various steps of sintering, such as grain growth in the initial step occurs due to evaporation, condensation and surface diffusion, however, the grain boundary diffusion leads to an improvement in density; grain growth or densification can also occur through lattice diffusion subject to the origin of material transfer [1], though the extent of the grain boundary diffusion depends on the available grain boundary area or the propensity of the grain growth. In the intermediate stage, the contraction in pore diameter can occur either by diffusion through the lattice or the grain boundary [2]. The lattice diffusion-based rate of densification is [1]:

$$\frac{1}{\rho}\left(\frac{d\rho}{dt}\right) = \frac{10 D_1 \gamma_{sv} \Omega}{\rho G^3 kT} \tag{6.1}$$

where $\frac{1}{\rho}\left(\frac{d\rho}{dt}\right)$ is the densification rate, D_1 is the lattice diffusion coefficient, Ω is the molecular volume, G is the grain size, k is the Boltzmann constant, γ_{sv} is the surface energy, r is the pore radius, and T is the absolute temperature.

The rate of densification $\left[\frac{1}{\rho}\left(\frac{d\rho}{dt}\right)\right]$ by grain boundary diffusion is [1]:

$$\frac{1}{\rho}\left(\frac{d\rho}{dt}\right) = \frac{4}{3}\left[\frac{D_{gb} \delta_{gb} \gamma_{sv} \Omega}{\rho(1-\rho)^{1/2} G^4 kT}\right] \tag{6.2}$$

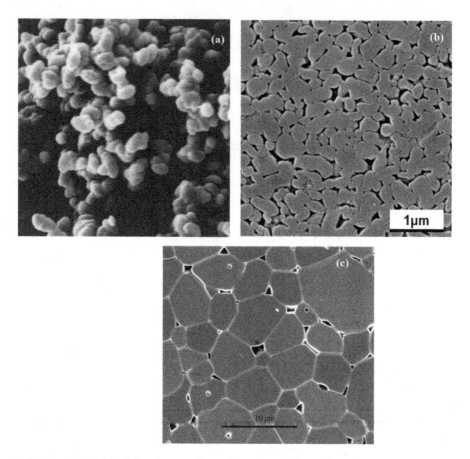

FIGURE 6.1 (a) Initial sintering step shows the neck attaching with particles, (b) intermediate sintering step evidences neck growth and pore network, (c) final sintering step presents different shapes of isolated pores normally present at the grain boundaries, different pore shape with varying locations are observed. (Reprinted from [1]. Copyright (2021), with permission from Elsevier.)

where D_{gb} is the diffusion coefficient of grain boundary, δ_{gb} is the grain boundary thickness.

The final step of densification includes the pore shrinkage and in the case of pore separation, the pore shape is controlled by the surface diffusion. Grain boundary diffusion and lattice diffusion regulate the pore shrinkage mechanism [1]. The pore shrinkage depends on the dihedral angle (the angle between the grain boundary and the pore surface) and, for the grain boundary pore, the dihedral angle is related to the grain boundary energy (γ_b): the surface energy (γ_s) as [1]:

$$\frac{\gamma_b}{\gamma_s} = 2\cos\left(\frac{\psi}{2}\right) \quad (6.3)$$

The pore stability also depends on the dihedral angle, as a less dihedral angle leads to stabilization of the pore and the pore shrinkage needs a higher driving force with an increase in the dihedral angle. The increase in grain size also causes pore shrinkage, though the rate of shrinkage is less with higher grain growth [1]. Higher γ_b/γ_s and less dihedral angle are related to pores with a high aspect ratio whereas lower γ_b/γ_s and a high dihedral angle lead to developing spherical pores [1].

The sintering process can be accelerated with a reduced sintering temperature by the liquid phase sintering method [3]. Liquid phase sintering involves the formation of a liquid phase in which the solid constituents are soluble. Therefore, the effectiveness of the liquid phase depends on the solubility of the solid constituents in the liquid. The wetting of the solid phase by the liquid and the capillary pressure exerted by the liquid phase are effective in particle bonding. If the solid has no solubility in liquid, the solid-state sintering is dominant and the liquid assists in densification by sealing the pores [4, 5]. The liquid-phase sintering process and the progress of densification with sintering time are described in Figure 6.2 and Figure 6.3 respectively. In the initial step, the solid state sintering occurs which leads to densification. With the progress of sintering, the liquid phase moves between the grains and grain readjustment occurs, followed by solution-reprecipitation and finally solid state sintering, which induces slow densification [6].

The liquid-phase sintering depends on the wettability of the liquid on the solid surface and therefore the wetting angle, as displayed in Figure 6.4 influences the densification behavior. The contact or wetting angle (θ) is related to the solid-vapor interfacial energy (γ_{SV}), the solid-liquid interfacial energy (γ_{SL}), and the liquid-vapor interfacial energy (γ_{LV}) as [6]:

$$\gamma_{SV} = \gamma_{SL} + \gamma_{LV} \cos(\theta) \tag{6.4}$$

If the wetting angle is high (poor wetting), the densification is reduced as the liquid discharges from the pores, whereas at a smaller wetting angle, the capillary pressure of the liquid induces densification.

The dihedral angle (ϕ) also regulates the liquid phase sintering and is associated with the surface energy of the solid-liquid (γ_{SL}) and the grain boundary energy (γ_{SS}) according to the following equation [6]:

$$2\gamma_{SL} \cos\left(\frac{\phi}{2}\right) = \gamma_{SS} \tag{6.5}$$

The liquid enters the grain boundaries which are solid when the surface energy ratio between solid-solid and solid-liquid is > 1.8 and ϕ tends to 0° and the condition facilitates liquid-phase sintering [6]. Reports indicate the liquid-phase sintering of W-Cu, Mo-Cu, though the solubility of W and Mo in Cu is less, which can be counteracted by particle size refinement prior to sintering or the addition of a sintering activator such as Ni, Pd, Co, Fe in the W-Cu system [7–10]. An activator for liquid-phase sintering is needed to reduce the activation energy and also delivers a short-circuit channel for the material transfer. The effectiveness of the sintering activator is associated with several factors: (1) the grain boundary segregation; (2) the

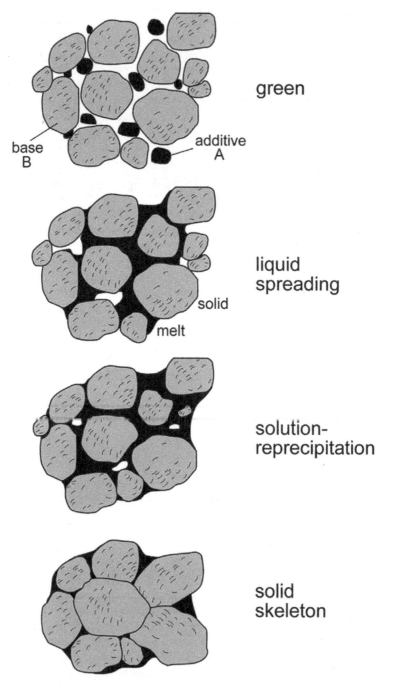

FIGURE 6.2 Microstructural variation during liquid-phase sintering. (Reprinted from [6]. Copyright (2014), with permission from Elsevier.)

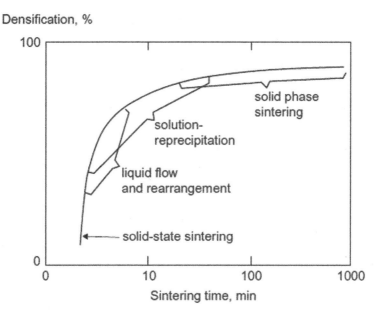

FIGURE 6.3 Liquid-phase sintering process: variation of percentage densification with sintering time. (Reprinted from [6]. Copyright (2014), with permission from Elsevier.)

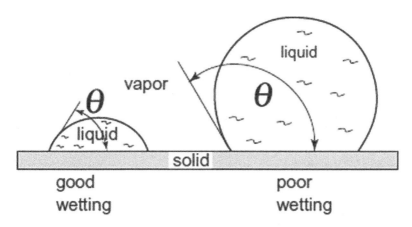

FIGURE 6.4 Wetting characteristics based on wetting angle. (Reprinted from [6]. Copyright (2014), with permission from Elsevier.)

reduction in liquidus and solidus temperature along with the higher addition of additive; (3) the additive being ductile; (4) the high enthalpy of sublimation; and (5) the smaller size of the atom [11–14]. A densification study of W-Ni-2.0wt.%Y_2O_3, W-Ni-2.0wt.%ZrO_2, W-Ni-2.0wt.%TiO_2 indicates a higher sintered density in W-Ni-2.0wt.%Y_2O_3 alloys against other alloys due to the prevention of oxide formation due to the presence of oxygen in the milled powders, whereas for other alloys, the formation of gas bubbles resulting from the incipient oxide melting leads to an increase in the porosity [15]. The densification in Y_2O_3 dispersed W alloys is also allied with the presence of solid-phase sintering (due to the existence of polyhedral grains) and liquid-phase sintering if the sintering temperature is higher than the eutectic temperature of W-Y_2O_3 (1560°C) [16] or $Y_2(WO_4)_3$ (melting temperature: 1440°C) [17]. The densification of sintered W-Y_2O_3 alloys increases with a sintering temperature between 1800–2500°C, however, the increment is steep till 2000°C, followed by shallow enhancement in percentage relative density [18]. The percentage relative densities of sintered W-(0.5, 1.0, 1.5, 2.0 wt.%)Y_2O_3 alloys at 1800°C are less than 80% and the densification improves beyond 90% and less than 99% at 2500°C [18]. Other papers also evidence the application of sintering temperature of 1800°C (2073 K) for W-20vol.% Y_2O_3 alloys to achieve percentage relative density in excess of 99% [19]. The external pressure application during sintering provides the additional driving force. The pressure applied (P_a) is related to the actual operative pressure (ΔP_e) as [1]:

$$\Delta P_e = \phi P_a \tag{6.6}$$

where ϕ is the factor of stress intensification. The operative pressure also includes surface energy (γ) and the radius of curvature (R) and the corresponding relationship will be:

$$\Delta P = \frac{2\gamma}{R} + \phi P_a \tag{6.7}$$

The higher surface energy contributes to increasing the operative pressure. Higher pressure application reduces the temperature gradient during SPS and improves the compactness of the die-punch and conduction [20]. The prime rate regulating mechanism for W sintered by SPS within 1250–1500°C with pressure < 16 MPa is the grain boundary diffusion to the particle neck area in the initial stage [21]. Furthermore, it is reported that the rate regulating stage within the temperature range of 0.4–0.65 T_m and > 0.65 T_m is the stage for dislocation diffusion and lattice diffusion, respectively (T_m: W melting temperature) [22]. The steps during SPS of W-10Ti have been described by Huang et al. [23]. According to the paper, during the first step (temperature: < 800°C), application of pressure incorporates plastic deformation, which contributes to densification, the second step (temperature: 800–1500°C) comprises the development and growth of the neck, which leads to volume shrinkage and enhancement of densification, however, after a specific temperature the degree of shrinkage decreases. The third stage (temperature greater than 1500°C) includes a reduced densification rate owing to the enhancement in diffusion length and the shrinkage which achieve constancy [23]. The strain rate sensitivity (m) is also related to the operative sintering mechanism during the hot pressing of W [24, 25]. If the m value is in the range of 0.2–0.25. the working sintering mechanism is a diffusion-assisted dislocation climb at the last step of sintering [24, 25].

6.2 STRATEGY TO ENHANCE DENSIFICATION

Section 6.1 discusses different operative mechanisms during densification. The final consolidated density depends on several factors in powder metallurgy, such as particle size, shape, compaction pressure, type of compaction, load application rate and residual time during compaction, mode of sintering (solid/liquid phase), sintering temperature, heating rate, sintering time and atmosphere, applied pressure during sintering, and the concentration of oxide dispersion. In pressureless sintering, it is of prime significance to achieve superior green density to improve the sintered density. Increase in compaction pressure improves the density in several steps, such as: (1) particle rearrangement; (2) enhancement in particle coordination number and contact area, which is related to elastic deformation; (3) plastic deformation; and (4) bulk deformation [26]. The effect of compaction pressure on the green density of the compact is described by Panelli's equation as [27]:

$$\ln\left(\frac{1}{1-D}\right) = A_0 P^{1/2} + B_0 \tag{6.8}$$

where D is the relative density of compact, P is the applied pressure, A_0 is the plastic deformation capacity of powders, B_0 is the density without pressure application.

The higher A_0 and B_0, the higher will be the density. The A_0 can be enhanced by proper alloy design or selection of ductile constituents; however, in the case of ductile-brittle powder, B_0 can be enhanced by coarse: fine particles. Fine particles can fill the void space between coarser particles and improve particle packing. Spherical and coarser particles also flow more easily in the die with reduced interparticle friction than elongated and irregular shaped particles. The phenomenon also contributes to superior compressibility. If the height to diameter ratio of compacted pellets is higher, the density gradient will increase in uniaxial compaction. Adopting isostatic pressing facilitates a reduction in the density variation in the compact which can produce isotropic properties. Many anomalies lead to poor compaction of powders during uniaxial compaction as described below [28]:

1. Off-centering of punch-die set with respect to load application axis of compaction press.
2. Manufacturing defect of die-punch (inadequate tolerance).
3. Bonding of powder at the inner surface of die.
4. Air encapsulation by powder during transport.

The above factors need to be controlled and maintained to achieve a better compactability and prevent a broken edge of compacts.

The density can also be further improved by compacting under warm conditions. The load application rate and the residual time after maximum load is attained need to be controlled throughout the compaction. These factors are more significant if the compact dimension is higher. A study presents that with an increase in percentage relative green density at 450°C compaction temperature, compared to room temperature, compaction is more significant beyond 400 MPa pressure, and maximum percentage relative green density of 58% has been achieved at 450°C and 800 MPa pressure [29]. The improvement in percentage relative green density at 450°C is attributed to the enhanced plastic deformation ability of the powder particles [29]. The paper also illustrates that higher percentage relative green density (55.1%) increases the percentage relative sintered density (98.1%) when sintering is carried out in a hydrogen atmosphere at 1100°C for 1 h [29]. The presence of the liquid phase, as described in Section 6.1, assists in densification during sintering. Sintering of W-Ni-Fe by hot pressing (fixed pressure) and by oscillating pressure sintering within 900–1350°C shows that increasing the sintering temperature improves the percentage relative sintered density [30]. The percentage sintered density attained by oscillating pressure sintering is 99% with a corresponding sintering temperature at 1250°C, however, the same densification is recorded by hot pressing at 1350°C and it is anticipated that plastic deformation and grain boundary sliding contribute to superior densification in oscillating pressure sintering [30]. Apart from oscillating pressure sintering, the laser additive manufacturing technique of W can also develop high sintered density in much less time than traditional liquid phase sintering [31]. Higher heating rate during sintering also reduces the percentage relative sintered density owing to significant deviation in temperature gradient between the inner and the outer sections of the sintered product [32]. Yao et al. have reported that increasing the heating rate from 77°C/min to 125°C/min reduces the percentage relative sintered density from 98.57% to 95.45% at 1700°C [32]. Variation of sintering time (1, 3, 5, 7, 9 min) at 30 MPa pressure at maximum temperature of 975°C for W-Mo-Cu composites prepared by electric field sintering reports that an increase in sintering time increases the percentage relative sintered density till 5 min to 94.84%, however, further enhancement in sintering time to 9 min reduces the percentage densification to 93.43%, owing to the presence of higher diameter pores [33]. Lee et al. have described how the increase in holding time from 0 to 30 min at a constant temperature of 1600°C, at pressure of 60 MPa, during spark plasma sintering of W in a vacuum increases the percentage relative density from 81.18% to 95.93% [34]. A similar trend is also reported in spark plasma sintered NbTi-Ta alloys and the reduced sintered density at the lower holding time is due to inadequate neck formation [35]. The inherent pressure effect on sintered density depends on the particle size as a reduced particle size, the influence of pressure is less on densification and it can be increased by an elevated pressure level [36]. Experiments show the effect of applied pressure during field-assisted sintering of W-11.4Cr-0.6Y-0.4Zr alloy at 1400°C, which shows that increasing pressure from 10 MPa to 60 MPa enhances the percentage relative sintered density from 84.4 ± 0.3 % to 99.5 ± 0.3 % respectively due to the superior displacement-rearrangement and plastic deformation mechanism [37]. Another study on the 3D multi-particle FEM method shows that increasing the HIP pressure from 0 MPa to 300 MPa at a constant temperature also improves the density, however, the improvement is nominal between 200–300 MPa pressure (1.27%) against the significant enhancement of 91.1% amid 0–100 MPa [38]. The selection of sintering atmosphere is vital to control the final densification, pore size and mechanical properties. Several studies indicate that a dry hydrogen atmosphere during sintering of W heavy alloys results in swelling and blistering, which negatively impact the sintered density, and the phenomenon can be counteracted by a dry followed by wet H_2 atmosphere [39, 40]. Further embrittlement and extended deviation in mechanical properties of alloys sintered in an H_2 environment impose challenges, and these can be fixed by: (1) adopting an argon (Ar) atmosphere in the last phase of sintering; and (2) additional heat treatment after sintering [39]. During liquid-phase sintering the solubility and diffusivity of gas in the liquid phase determine the extent of densification [39]. It is reported that several gases, such as argon, or water vapor, which possesses limited solubility inside the liquid phase, get confined in the pores and inhibit full densification, whereas a vacuum atmosphere leads to improved densification due to no gas sealing in the pores [39]. Reports describe how pore size can collapse to undetectable with the increase in sintering time in a vacuum compared to complete Ar or H_2 atmosphere [41]. The effect of oxide dispersion on sintered density has a different trend while varying the concentration. An experimental study suggests that an increase in oxide dispersion leads to improvement in the densification up to a critical content beyond which the densification drops. The addition of ZrO_2 in W results in enhancement of densification, but the extent of increment is less at 2.0 wt.% [42]. The grain growth velocity also impacts the density, as restricting the grain growth can enrich the densification [43], whereas the sintering can be triggered by the stored internal energy during the ball milling of the alloys which further enhance the sintered density [44]. The energy storage can be increased by the addition of higher nano-oxides in refractory metals employed in mechanical alloying. Zhou et al. have varied the Al_2O_3 from 4.15 wt.% to 20.60 wt.% in $Mo-Al_2O_3$ composites and report that the density of the composite decreases with an increase in the Al_2O_3 content [45]. Increasing the La_2O_3 content from 0.5 wt.% to 5.0 wt.% decreases the percentage relative sintered density of $W-La_2O_3$ alloys due to the restriction in the neck formation and the growth required for densification [46]. Another paper describes how impeding diffusion by oxide dispersion reduces the percentage relative density of W-0.9 wt.% La_2O_3 (89%) compared to pure W (91%) [47]. Wang et al. have proposed that increasing Lu_2O_3 wt.% from 1% to 3% can enrich the percentage densification from 89.37% to 96.21% respectively, though the percentage relative density of 1wt.% and 2 wt.% is less against pure W [48]. Contrarily, Sm_2O_3 addition in W-1 wt% Sm_2O_3 alloy consolidated by SPS has been reported to improve the percentage relative density (97.8%) against pure W (95.8%) [49]. Kim et al. have compared the percentage densification between $W-Y_2O_3$, $W-HfO_2$ and $W-La_2O_3$ composites fabricated by SPS at 1700°C for 3 min [50]. The paper presents that maximum percentage relative density is attained for $W-HfO_2$ followed by $W-Y_2O_3$ and $W-La_2O_3$ with 0.5 wt.% oxide addition, though the trend changes at higher oxide wt.% and maximum percentage relative density of almost 100% is reported for $W-Y_2O_3$ with 5 wt.% Y_2O_3 addition [51].

6.3 ROLE OF DENSIFICATION IN FINAL APPLICATIONS

The applications of refractory alloys are extensive and every application has specific property requirements. The application of refractory alloys in high temperature applications, such as in aerospace, requires resistance against oxidation as well as sufficient high temperature strength. During oxidation of some refractory metals (W, Mo), the volatile constituents get removed as well as oxygen diffusion occurs through short-circuiting channels. The presence of porosities aggravates the oxidation further and may lead to premature failure of components. Therefore, superior densification reduces the oxidation as well as positively contributing to strength. Porosity facilitates crack nucleation and therefore enhancement of the impact toughness of W by increasing the density is a essential approach [51]. A report also suggests applying swagging and rolling as fabrication methods to improve the fracture strength [51]. During neutron irradiation in a nuclear reactor, both nanostructure grains and ductility are significant. High relative density (around 99.5%) is associated with the ductilization effect in refractory alloys [51]. The recent literature describes how through computational investigation the presence of compacted nanoparticles (52% densification with respect to bulk W) can be effective as plasma-facing material, owing to the rapid movement of point defects to the surface and eradication [52]. Superior density effectively inhibits radiation from high energy α and γ particles in a fusion reactor [53]. For a kinetic energy penetrator application, high density is an important aspect in achieving superior penetration depth. Fang et al. have studied the penetration performance between W alloys (90W–7Ni–3Fe and 80W–14Cu–6Zn) and indicate that the superior density (17 g/cm^3) in 90W–7Ni–3Fe alloy reduces the impact velocity and penetration depth against 80W–14Cu–6Zn (density: 15 g/cm^3), and specify that additionally self-sharpening behavior, quick plastic localization, and adiabatic shear band formation add to the improved penetration performance [54, 55]. Complete densification is also impervious to erosion and cracking in the application of refractory metal in a rocket nozzle, though the effectiveness depends on the flame temperature of the propellants [56]. W nozzles, fabricated by the powder metallurgy route with less density, erode through chemical reaction and mechanical abrasion mode and, conversely, W with a high density exhibits minor erosion through oxidation [56]. The presence of extensive voids or cavities may assist in reduced abrasion by facilitating crack propagation. W 25%Re alloy, developed by powder injection molding (PIM) followed by pressureless sintering and HIP for a rocket nozzle application, exhibits final percentage relative density of 99% and shows exceptional resistance to erosion (0.009 mm/s) [57].

Some other applications, such as aviation counterweights or tool components, also need to be prepared with high density. The corrosion resistance of Nb-Ag alloys can also be improved by increasing the density due to the formation of a continuous impervious layer, as evident from the study carried out by Wan et al. [58]. The reported current density of corrosion for Nb, Nb-1 at.% Ag, Nb-5 at.% Ag is 5.15×10^{-6} A/cm^2, 6.46×10^{-7} A/cm^2, 2.26×10^{-7} A/cm^2 respectively, corresponding to the percentage relative density 91.8%, 95.1% and 98.8% respectively [58]. The corrosion-resistant alloys are useful for biomedical applications owing to their biocompatibility, appreciable strength and wear resistance [58]. The above examples show that enriching densification is highly effective in improving resistance to wear/abrasion, oxidation resistance, and ductility in high temperature structural applications. Although increasing the sintering temperature improves the sintered density but with related shrinkage. The shrinkage is higher when liquid-phase sintering is accomplished and the final dimension of the sintered product is distorted. Therefore, the dimension of the compacted product needs to be increased where a higher sintering time is employed [59, 60].

6.4 INFERENCES

Increasing densification prior to sintering in the pressureless sintering technique and during sintering in pressure-based and pressureless sintering has an immense influence on improving the mechanical properties and high temperature sustainability (oxidation resistance). As discussed above, the factors affecting densification during synthesis (particle size, particle shape, impurities) and consolidation (compaction pressure, mode of compaction: uniaxial/isostatic, sintering parameters) need to be judiciously controlled to improve the uniformity in properties. The densification can be enhanced by increasing the temperature, though at the cost of coarsened grains and increased cost which can be addressed by increasing the heating rate such as in spark plasma sintering or the microwave sintering method. An attempt to improve density after consolidation by pressureless conventional sintering by the addition of a sintering aid (liquid-phase sintering) is associated with shrinkage due to the arrangement of the grain shape. The shrinkage can be reduced with initial coarser particles. The swelling tendency during sintering is also impacted by the solubility of the solid phase in the sintering aid and higher solubility increases the swelling tendency [3]. The limitations of SPS regarding industrial-scale production are the sample dimension, reduced yield and challenges in complex shape development [61]. Conventional sintering and powder forging methods are mainly used to fabricate powder metallurgical parts.

REFERENCES

[1] J. E. Blendell, W. Rheinheimer, Solid-state sintering, in: M. Pomeroy (Ed.), *Encyclopedia of Materials: Technical Ceramics and Glasses*, Elsevier, Oxford, vol. 1, (2021) 249–257.
[2] M. N. Rahaman, *Sintering of Ceramics*, CRC Press: Boca Raton, FL, (2007).
[3] R. M. German, P. Suri, S. J. Park, Review: Liquid phase sintering. *J. Mater. Sci.*, 44 (2009) 1–39.
[4] J. L. Johnson, R. M. German, Solid-state contributions to densification during liquid-phase sintering, *Metall. Mater. Trans. B*, 27 (1996) 901–909.
[5] J. L. Johnson, R. M. German, Role of solid-state skeletal sintering during processing of Mo-Cu composites, *Metall. Mater. Trans. A*, 32 (2001) 605–613.
[6] R. M. German, Sintering with a liquid phase, in: R. M. German (Ed.), *Sintering: From Empirical Observations to Scientific Principles*, Butterworth-Heinemann, Oxford, (2014) 247–303.
[7] J. L. Johnson, Activated liquid phase sintering of W-Cu and Mo-Cu, *Int. J. Refract. Met. Hard Mater.*, 53 (Part B) (2015) 80–86.
[8] J. H. Yu, T. H. Kim, J. S. Lee, Particle growth during liquid phase sintering of nanocomposite W-Cu powder, *Nanostruct. Mater.*, 9 (1997) 229–232.
[9] J. L. Johnson, Grain growth during liquid phase sintering of W-Cu, in: M. Bulger, B. Stebick (Eds.), *Advances in Powder Metallurgy and Particulate Materials*, Metal Powder Industries Federation, Princeton, NJ, (2010), 61–72.
[10] J. L. Johnson, R. M. German, Phase equilibria effects on the enhanced liquid phase sintering of tungsten-copper, *Metall. Mater. Trans. A*, 24 (1993) 2369–2377.
[11] J. J. Burton, E. S. Machlin, Prediction of segregation to alloy surfaces from bulk phase diagrams, *Phys. Rev. Lett.*, 37 (1976) 1433–1436.

[12] M. P. Seah, Adsorption induced interface cohesion, *Acta Metall.*, 28 (1980) 955–962.
[13] J. P. Stark, H. L. Marcus, The influence of segregation on grain boundary cohesion, *Met. Trans.*, 8A (1977) 1423–1429.
[14] E. D. Hondros, M. P. Seah, Segregation to interfaces, *Inter. Metals Rev.*, 22 (1977) 262–301.
[15] W. M. R. Daoush, A. H. A. Elsayed, O. A. G. E. Kady, M. A. Sayed, O. M. Dawood, Enhancement of physical and mechanical properties of oxide dispersion-strengthened tungsten heavy alloys. *Metall. Mater. Trans. A*, 47 (2016) 2387–2395.
[16] O. N. Carlson, The O-Y (oxygen–yttrium) system, *Bull. Alloy Phase Diagrams*, 11 (1) (1990) 61–66.
[17] M. N. A. Fenoel, R. Taillard, J. Dhers, J. Foct, Effect of ball milling parameters on the microstructure of W-Y powders and sintered samples, *Int. J. Refract. Met. Hard Mater.*, 21 (3–4) (2003) 205–213.
[18] Y. Kim, M. H. Hong, S. H. Lee, E. P. Kim, S. Lee, J. W. Noh, The effect of yttrium oxide on the sintering behavior and hardness of tungsten, *Met. Mater. Int.*, 12 (2006) 245–248.
[19] Y. Itoh, Y. Ishiwata, Strength properties of yttrium-oxide-dispersed tungsten alloy, *JSME Int. J. Series*, 39 (3) (1996) 429–434.
[20] S. Grasso, Y. Sakka, G. Maizza, Pressure effects on temperature distribution during spark plasma sintering with graphite sample, *Mater. Trans.*, 50 (8) (2009) 2111–2114.
[21] Z. Gao, G. Viola, B. Milsom, Kinetics of densification and grain growth of pure tungsten during spark plasma sintering, *Metall. Mater. Trans. B*, 43 (2012) 1608–1614.
[22] S. L. Robinson, O.D. Sherby, Mechanical behavior of polycrystalline tungsten at elevated temperature, *Acta Metall.*, 17 (1969) 109–125.
[23] L. Huang, J. Zhang, Y. Pan, Y. Du, Spark plasma sintering of W-10Ti high-purity sputtering target: Densification mechanism and microstructure evolution, *Int. J. Refract. Met. Hard Mater.*, 92 (2020) 105313.
[24] D. M. Karpinos, A. A. Kravchenko, Y. L. Pilipovskii, V. G. Tkachenko, Y. M. Shamatov, Hot pressing of tungsten and its pseudo alloys. Part I, Sov., *Powder Metall. Met. Ceram.*, 9 (1970) 287–291.
[25] D. M. Karpinos, A. A. Kravchenko, Y. L. Pilipovskii, V. G. Tkachenko, Y. M. Shamatov, Hot pressing of tungsten and its pseudo alloys. Part II, Sov., *Powder Metall. Met. Ceram.*, 10 (1971) 367–372.
[26] R. M. German, *Powder Metallurgy Science*, (2nd edn.), Metal Powder Industries Federation, Princeton, NJ, (1994).
[27] R. Panelli, F. A. Filho, Compaction equation and its use to describe powder consolidation behavior, *Powder Metall.*, 41 (1998) 131–133.
[28] A. F. M. A. Siddiqui, A. Patra, General study, National Institute of Technology, Rourkela, (2021).
[29] X. Wang, Z. Z. Fang, M. Koopman, The relationship between the green density and as-sintered density of nano-tungsten compacts, *Int. J. Refract. Met. Hard Mater.*, 53 (Part B) (2015) 134–138.
[30] K. Gao, Y. Xu, G. Tang, L. Fan, R. Zhang, L. An, Oscillating pressure sintering of W-Ni-Fe refractory alloy, *J. Alloys Compd.*, 805 (2019) 789–793.
[31] S. Zhou, Y. J. Liang, Y. Zhu, B. Wang, L. Wang, Y. Xue, Ultrashort-time liquid phase sintering of high-performance fine-grain tungsten heavy alloys by laser additive manufacturing, *J. Mater. Sci. Technol.*, 90 (2021) 30–36.
[32] L. Yao, Y. Gao, Y. Huang, Y. Li, X. Huang, P. Xiao, Fabrication of Mo-Y_2O_3 alloys using hydrothermal synthesis and spark plasma sintering, *Int. J. Refract. Met. Hard Mater.*, 98 (2021) 105558.
[33] H. L. Zhou, K. Q. Feng, S. X. Ke, Y. F. Liu, Densification and properties investigation of W-Mo-Cu composites prepared by large current electric field sintering with different technologic parameter, *J. Alloys Compd.*, 767 (2018) 567–574.

[34] G. Lee, J. McKittrick, E. Ivanov, E. A. Olevsky, Densification mechanism and mechanical properties of tungsten powder consolidated by spark plasma sintering, *Int. J. Refract. Met. Hard Mater.*, 61 (2016) 22–29.

[35] Q. Wang, G. Cui, H. Chen, Effect of the Ta addition on densification and mechanical properties of NbTi alloys prepared by SPS, *J. Alloys Compd.*, 868 (2021) 159106.

[36] Z. A. Munir, U. A. Tamburini, M. Ohyanagi, The effect of electric field and pressure on the synthesis and consolidation of materials: A review of the spark plasma sintering method, *J. Mater. Sci.*, 41 (2006) 763–777.

[37] S. P. Yang, et al., Influence of the applied pressure on the microstructure evolution of W-Cr-Y-Zr alloys during the FAST process, *Fusion Eng. Des.*, 169 (2021) 112474.

[38] Y. Zou, X. An, R. Zou, Investigation of densification behavior of tungsten powders during hot isostatic pressing with a 3D multi-particle FEM approach, *Powder Technol.*, 361 (2020) 297–305.

[39] A. Bose, R. M. German, Sintering atmosphere effects on tensile properties of heavy alloys, *Metall. Mater. Trans. A*, 19 (1988) 2467–2476.

[40] R. M. German, A. Bose, S. S. Mani, Sintering time and atmosphere influences on the microstructure and mechanical properties of tungsten heavy alloys, *Metall. Mater. Trans. A*, 23 (1992) 211–219.

[41] R. M. German, K. S. Churn, Sintering atmosphere effects on the ductility of W-Ni-Fe heavy metals, *Metall. Mater. Trans. A*, 15 (1984) 747–754.

[42] C. Wang, L. Zhang, S. Wei, K. Pan, X. Wu, Q. Li, Effect of ZrO_2 content on microstructure and mechanical properties of W alloys fabricated by spark plasma sintering, *Int. J. Refract. Met. Hard Mater.*, 79 (2019) 79–89.

[43] M. Zhao, Z. Zhou, Q. Ding, M. Zhong, K. Arshad, Effect of rare earth elements on the consolidation behavior and microstructure of tungsten alloys, *Int. J. Refract. Met. Hard Mater.*, 48 (2015) 19–23.

[44] Y. Han, J. Fan, T. Liu, H. Cheng, J. Tian, The effects of ball-milling treatment on the densification behavior of ultra-fine tungsten powder, *Int. J. Refract. Met. Hard Mater.*, 29 (6) (2011) 743–750.

[45] Y. Zhou, Y. Gao, S. Wei, K. Pan, Y. Hu, Preparation and characterization of Mo/Al_2O_3 composites, *Int. J. Refract. Met. Hard Mater.*, 54 (2016) 186–195.

[46] Z. Chen, J. Yang, L. Zhang, B. Jia, X. Qu, M. Qin, Effect of La_2O_3 content on the densification, microstructure and mechanical property of $W-La_2O_3$ alloy via pressureless sintering, *Mater. Charact.*, 175 (2021) 111092.

[47] M. A. Yar, S. Wahlberg, H. Bergqvist, H. G. Salem, M. Johnsson, M. Muhammed, Chemically produced nanostructured ODS–lanthanum oxide–tungsten composites sintered by spark plasma, *J. Nucl. Mater.*, 408 (2) (2011) 129–135.

[48] S. Wang, J. Zhang, L-M. Luo, X. Zan, Q. Xu, X-Y. Zhu, K. Tokunaga, Y. C. Wu, Properties of Lu_2O_3 doped tungsten and thermal shock performance, *Powder Technol.*, 301 (2016) 65–69.

[49] X-Y. Ding, L-M. Luo, X-Y. Tan, G-N. Luo, P. Li, X. Zan, J-G. Cheng, Y-C. Wu, Microstructure and properties of tungsten–samarium oxide composite prepared by a novel wet chemical method and spark plasma sintering, *Fusion Eng. Des.*, 89 (6) (2014) 787–792.

[50] Y. Kim, K. H. Lee, E.P. Kim, D. I. Cheong, S. H. Hong, Fabrication of high temperature oxides dispersion strengthened tungsten composites by spark plasma sintering process, *Int. J. Refract. Met. Hard Mater.*, 27 (5) (2009) 842–846.

[51] C. Linsmeier, et al., Development of advanced high heat flux and plasma-facing materials, *Nucl. Fusion*, 57 (2017) 092007.

[52] P. D. Rodriguez, F. Munoz, J. Rogan, I. M. Bragado, J. M. Perlado, O. P. Rodriguez, A. Rivera, F. J. Valencia, Highly porous tungsten for plasma-facing applications in

nuclear fusion power plants: a computational analysis of hollow nanoparticles, *Nucl. Fusion*, 60 (9) (2020) 096017.

[53] Y. Sahin, Recent progress in processing of tungsten heavy alloys, *J. Powder Technol.*, 2014, 764306.

[54] X. Fang, J. Liu, X. Wang, S. Li, W. Guo, Investigation on the penetration performance and "self-sharpening" behavior of the 80W-14Cu-6Zn penetrators, *Int. J. Refract. Met. Hard Mater.*, 54 (2016) 237–243.

[55] L. S. Magness, High strain rate deformation behaviors of kinetic energy penetrator materials during ballistic impact, *Mech. Mater.*, 17 (2–3) (1994) 147–154.

[56] J. R. Johnston, R. A. Signorelli, J. C. Freche, Performance of rocket nozzle materials with several solid propellants, *NASA Technical Note*, NASA TN D-3428, (1966).

[57] D. Y. Park, Y. J. Oh, Y. S. Kwon, S. T. Lim, S. J. Park, Development of non-eroding rocket nozzle throat for ultra-high temperature environment, *Int. J. Refract. Met. Hard Mater.*, 42 (2014) 205–214.

[58] T. Wan, K. Chu, J. Fang, C. Zhong, Y. Zhang, X. Ge, Y. Ding, F. Ren, A high strength, wear and corrosion-resistant, antibacterial and biocompatible Nb-5 at.% Ag alloy for dental and orthopedic implants, *J. Mater. Sci. Technol.*, 80 (2021) 266–278.

[59] A. A. Tounsi, M. S. J. Hashmi, Effect of sintering temperature on the densification, shrinkage and compressive strength of stainless steel 300 series, *J. Mater. Process. Technol.*, 37 (1993) 551–557.

[60] S. J. L. Kang, K. H. Kim, D. N. Yoon, Densification and shrinkage during liquid-phase sintering, *J. Am. Ceram. Soc.*, 74 (1991) 425–427.

[61] C. Maniere, E. Nigito, L. Durand, A. Weibel, Y. Beynet, C. Estournes, Spark plasma sintering and complex shapes: The deformed interfaces approach, *Powder Technol.*, 320 (2017) 340–345.

7 Mechanical and Wear Behavior

7.1 HARDNESS

The extent of hardness of sintered products is governed by both powder (size, size distribution, impurity content), compact properties (density) and sintering factors (time, temperature, composition). Any factor which increases the densification further improves the hardness of the matrix. Although improved density can be attained by increasing the sintering temperature, it counteracts the enhancement of hardness due to elevated grain growth. Therefore, an optimized density and grain size reflect the maximum hardness value. The hardness values are also influenced by the fabrication method. Several studies report the Vickers hardness of W-1wt.%Y_2O_3 at different sintering processes and temperature [1–4]. It is evident that the addition of 1 wt.% Y_2O_3 improves the Vickers hardness of W at the studied temperature (1500°C, 1800°C) and processing routes (spark plasma sintering, microwave sintering, hot pressing) [1–4]. The reported maximum hardness for W-1wt.%Y_2O_3 is 6.91 ± 0.20 GPa processed with microwave sintering at 1500°C sintering temperature [3]. In a comparison between the hardness of spark plasma and microwave sintered W-1wt.%Y_2O_3 alloy, though the microwave sintered alloy exhibits lower percentage relative density (96.8 ± 0.3 %), however, it possesses minimum grain size of 0.7 μm, which contributes to enhanced hardness [2, 3]. Processing of W-ZrO_2 alloy through hot isostatic pressing, hydrogen-based conventional sintering, vertical sintering, or hot swaging process shows that maximum microhardness of 486.3 HV is achieved by a hot isostatic pressing technique, though with reduced density and grain size [5]. The concentration of the oxide dispersion needs to be precisely controlled to achieve optimized density and hardness. An increase in wt.% of Al_2O_3 in W-Al_2O_3 improves the hardness of the alloy to 375 HV (with 0.25 wt.% Al_2O_3) and a further increase in wt.% of Al_2O_3 reduces the hardness [6]. The Vickers hardness is also impacted by applied load and loading time; therefore, depending on the grain size, the load must be judiciously selected. Another study on SPSed Mo-1wt.%Y_2O_3 illustrates that maximum microhardness of 314.77 ± 26.12 $HV_{0.3}$ has been attained with an optimized parameter of 1700°C sintering temperature, and a heating rate of 100°C/min [7]. The fine oxide particles interact with the dislocation, which contributes to an improvement in hardness. Improvement of hardness with an increase in wt.% of $Zr(Y)O_2$ and density is reported by Xiao et al. [8]. Maximum hardness of 486 HV is reported in the paper to correspond to 0.75 wt.% $Zr(Y)O_2$, grain size of 4.3 μm, oxide particle size of 0.38 μm and highest percentage relative density of 96.5% [8]. The hardness value is also stimulated by the formation of hard intermetallic phases apart from the presence of oxide particles. The highest hardness of 11.4 GPa is measured for W-4 wt.% SiC-5 wt.% Y_2O_3 owing to the higher presence of hard compounds (W_5Si_3, W_2C) and (W–Y) oxides [9]. Changing the processing condition also contributes to an improvement in hardness. Yar et al. have fabricated W-1%Y_2O_3 alloys using spark plasma sintering at much lower sintering temperatures of 1100°C and 1200°C and 3 min of holding time [10]. Changing the sintering cycle by introducing a holding time of 5 min at 900°C followed by holding for 1 min at the maximum sintering temperature of 1100°C enhances the hardness to 570 ± 30 HV_{200g} compared to a hardness of 518 ± 20 HV_{200g} at 1100°C with 3 min holding time [10]. The selection of a suitable oxide is also significant with respect to the improvement of hardness. Kim et al. have described how the addition of Y_2O_3 can potentially improve the hardness even at higher wt.% addition (5 wt.%) compared to HfO_2 and La_2O_3, and the hardness improvement by grain size reduction is not attributable to W-La_2O_3 but is influenced by the reduction in percentage relative density [11]. In a microhardness measurement study the concentration of the phase subjected to indentation results in the variation of hardness. Phases with a concentration of higher W content in W-based alloys improve the hardness. In another study, maximum hardness of 358.3 HV is achieved for W-Ni-TiO_2 alloy against W-Ni-ZrO_2 (333 HV) and W-Ni-Y_2O_3 (329.6 HV) alloys [12]. Although a higher percentage relative density is noted for W-Ni-Y_2O_3, the higher hardness in W-Ni-TiO_2 is due to finer grain size [12]. It is evident from the above discussion that the maximum hardness in ODS refractory alloys can be achieved by optimizing consolidation parameters, grain size, density and the concentration of oxide addition.

7.2 STRENGTH AND TOUGHNESS

Concurrent enrichment of strength and toughness is advantageous for structural materials. Higher strength with adequate toughness requires tailored microstructure before and after consolidation. Dispersion of oxides at the grain boundary refines the grains and improves the strength through a dispersion strengthening mechanism. The literature suggests that higher hardness does not always relate to higher strength [12]. Higher compressive strength in W-Ni-Y_2O_3 alloy (1565.4 MPa) corresponds to higher density and particle bonding even with a comparatively coarser microstructure than W-Ni-TiO_2 [12]. Refractory metals such as W, Mo, Re, Nb show high room temperature strength, but with an increase in temperature the strength reduces and therefore the addition of dispersoids of high temperature stability facilitates the preservation of adequate strength at high temperature. The major challenge for refractory metals (W, Mo) is inadequate room temperature percentage elongation. Several factors are responsible for the enhancement of strength and ductility by [13–16]:

1. Formation of a coherent lamellar grain boundary and reduction of boundaries' stress concentration.
2. Presence of deformation twins and increased high angle grain boundaries.
3. Dispersion of particles in nanoscale.
4. Bimodal grain size distribution other than monomodal grain size is beneficial to improve both the strength and percentage elongation.
5. Low porosities and impurities.

The boundary stress concentration can be relieved by dispersing the majority of the oxides inside the matrix rather than allowing complete dispersion at the boundaries. The mechanism combined with bimodal grain size distribution involves anchoring the dislocations inside the coarser matrix and further improving strain hardening [17]. The room temperature ultimate tensile strength and percentage total elongation in the longitudinal direction of ODS Mo are 746 ± 49 MPa and 14.1 ± 6.5%, respectively, and with an increase in temperature to 1000°C, the ultimate tensile strength and percentage total elongation value drop to 330 ± 23 MPa and 5.3 ± 0.6% respectively [18]. In another report, the tensile strength and percentage elongation of ODS-Mo (La_2O_3 dispersed Mo alloys) studied at different temperatures are 382 MPa, 10.7% (1050°C), 205 MPa, 8.7% (1200°C), 184 MPa, 11.4% (1300°C), 167 MPa, 9.6% (1400°C) respectively [17]. The report indicates the application window of ODS Mo alloys at high temperature. According to Gurwell, elongation and at least a fracture toughness of 6–8% and 33 MPa√m respectively are required for alloys used in a kinetic energy penetrator [19]. The requirement of high strength for an elevated heat effect in penetrator applications along with increased elongation can be attained by reducing the contiguity [20, 21]. The W/W grain boundary is quite weak, and increased contiguity can facilitate crack propagation. It is anticipated that reduced contiguity (detached grains) can decrease the mushroom-shaped formation at the penetrator's head and enrich the efficiency of the penetrator [22]. The yield strength (σ_y) is related to the volume fraction of the matrix (V_M) and the W particle diameter (D) in W alloys as [23]:

$$\sigma_y = \sigma_0 + k_2 Gb \left(\frac{1-V_M}{DV_M} \right)^{1/2} \quad (7.1)$$

where k_2 is the constant, G is the shear modulus, b is the Burgers vector. A decrease in particle size and volume fraction of the matrix increases the yield strength of the alloy [23].

The elongation to failure (ε) is also related to V_M and W/W contiguity (C_W) as [24]:

$$\varepsilon = \varepsilon_0 + k_2 V_M (1-C_W) \quad (7.2)$$

Equation (7.2) explains that reduced contiguity and the low volume fraction of matrix are the primary requirements to achieve augmented elongation to failure. Achievement of the best combination of strength and percentage elongation is related to dispersed oxide content. Dispersion of ZrO_2 in W alloys within the range of 0.5–2.0 wt.% exhibits a fluctuating trend with respect to ultimate compressive strength and percentage strain to failure [25]. Although the maximum densification is reported for W-2.0 wt.% ZrO_2 alloys, the highest room temperature compressive strength (1628 MPa) and strain to failure of 21% are measured for W-1.5 wt.% ZrO_2 [25]. Another strategy to enhance the strength and percentage elongation is the introduction of two-step mechanical alloying and two-step sintering (initially solid phase followed by liquid-phase sintering) [26]. However, the second-stage sintering temperature and time need to be maintained to establish a synergy between strength and percentage elongation. The high ductile-brittle transition temperature (DBTT) of W is related to inadequate ductility [27], and during annealing, the occurrence of the recrystallization process also increases the DBTT of W [28]. According to the study by Mabuchi et al., within a temperature of 1200–1700°C, the tensile strength improves and the percentage elongation to failure drops for rolled W-0.8 mass% La_2O_3 compared to pure W [29]. The higher stress exponent in W-La_2O_3 is dependent on the steadiness of the substructure provided by the oxide particles and the drop in elongation is accredited to the formation of cavities, as illustrated in the paper [29]. A study on HIP processed W-Y_2O_3 alloy points out the brittle behavior at room temperature, but the strength and toughness are enhanced with a temperature rise but drops beyond 600°C, though the values surpass pure W and W-Ti alloy [30]. The solubility of solute elements also influences the strength and toughness of an alloy. Counteracting the drop in ductility of ODS W alloys, a strategy such as the addition of Re, Ir in W has been investigated [31]. Higher Re addition (3.6 wt.%) is needed in W-1wt.%ThO_2 against lower Ir addition in W (0.4 wt.%) to achieve comparable ductility, though the strength of W-Re-ThO_2 is slightly higher than W-Ir alloy [31]. Higher Re also increases the cost and the increased addition (> 27 %) adds to the generation of intermetallic (σ phase) and depreciates the ductility [32]. The DBTT of Mo-ODS is lower (ranges with -25–75°C) compared to powder metallurgy processed Mo (around 50°C) in the recrystallized condition [33] and the results indicate the enrichment of ductility of Mo-ODS. The recrystallization process in refractory ODS alloys can be impeded by oxides at the grain boundary triple points which lower the tendency of grain boundary migration for recrystallization [34, 35].

7.3 MECHANISM OF DEFORMATION AT AMBIENT AND ELEVATED TEMPERATURES

The dispersion of oxides at the grain boundary may impact the deformation of the matrix owing to the differential deformation characteristics of the matrix and the oxide particles. The oxide particles facilitate stress concentration and assist in crack propagation, further reducing the ductility. W exhibits reduced thermal softening and slows down the flow softening at an elevated strain rate. The effects account for "mushrooming behavior" and increase the diameter of the kinetic energy penetrator (KEP), reducing the penetration depth. The requirement for shear banding, decreased diameter of penetrator and higher penetration depth aims to develop W heavy alloys with complete densification without any precipitates at the interface [21]. A dynamic deformation study indicates that the oxides at the interface can promote decohesion and encourage the adiabatic shear band formation, and a self-sharpening effect in W-Ni-Fe-Y_2O_3 [36]. The phenomenon indicates that W-ODS alloy is effective for KEP application. Oxide particles at the boundary can induce stress concentration and assist in intergranular crack formation, however, the literature shows that the source of the intergranular cracking during deformation of W-Y_2O_3 alloys at ambient temperature (25 °C) is the porosities [30]. The paper also describes how the boundary between W and Y rich regions has good cohesion without initiating cracks [30]. The temperature (T), flow stress (σ) effect on strain rate ($\dot{\varepsilon}$) is [37]:

$$\dot{\varepsilon} = A'\sigma^{n'} \exp\left(-\frac{Q}{RT}\right) \quad \text{(low stress)} \qquad (7.3)$$

$$\dot{\varepsilon} = A'' \exp(\beta\sigma) \exp\left(-\frac{Q}{RT}\right) \quad \text{(high stress)} \tag{7.4}$$

$$\dot{\varepsilon} = A[\sinh \alpha\sigma]^{n'} \exp\left(-\frac{Q}{RT}\right) \quad \text{(for every } \alpha\sigma\text{)} \tag{7.5}$$

where Q is the activation energy, $\dot{\varepsilon}$ is the strain rate (s^{-1}), R is the universal gas constant (8.314 J mol^{-1} K^{-1}); T is the absolute temperature, A, α, n' are constants without temperature dependency, $\beta = \alpha n'$.

True stress exhibits increasing trend with true strain, but the extent of increment is dependent on temperature and strain rate. The increase is not prominent and the flow stress achieves constancy at a low strain rate (0.0005 s^{-1}), and a decrease in flow stress is observed with increasing temperature of deformation from 1100°C to 1300°C of MoNbTaTiV refractory high entropy alloy as evident in Figure 7.1 (a–d) [38]. Enhancement of the temperature of deformation and the reduced strain rate transform the discontinuous dynamic recrystallization into a continuous dynamic recrystallization process, assisted by the presence of precipitate particles by the formation of a dislocation network [38].

High-temperature deformation behavior by simulation-based compression of W-0.25 wt.% Al$_2$O$_3$ alloy has been investigated by Wang et al. [39]. The authors report that true stress increases with increasing true strain, strain rate and a reduction in the temperature of study from 1300°C, 1400°C, 1500°C, 1600°C as presented in Figure 7.2 [39]. The work hardening mechanism is relaxed by the dynamic recovery/recrystallization process at a lower strain rate and a higher temperature, so dislocation easily slips, resulting in a drop in stress required for deformation [39].

The deformation (fixed temperature, strain rate) is provided by the constitutive equation as [39, 40]:

$$\sigma = K\dot{\varepsilon}^m \tag{7.6}$$

where σ is the flow stress, $\dot{\varepsilon}$ is the strain rate, m is the strain rate sensitivity exponent, K is the material constant.

m can be evaluated as:

$$m = \left[\frac{\partial \log \sigma}{\partial \log \dot{\varepsilon}}\right]_{\varepsilon,T} \approx \frac{\Delta (\log \sigma)}{\Delta (\log \dot{\varepsilon})} \tag{7.7}$$

The instability conditions is based on the following equation:

$$\xi(\dot{\varepsilon}) = \frac{\partial \log \frac{m}{m+1}}{\partial \log \dot{\varepsilon}} + m < 0 \tag{7.8}$$

The instability in the stress-strain diagram in Figure 7.2 during the hot condition leads to cracking, an adiabatic shear band and the lower instability is reported at lower strain, lower strain rate and a lower temperature regime [39]. An increase in the temperature

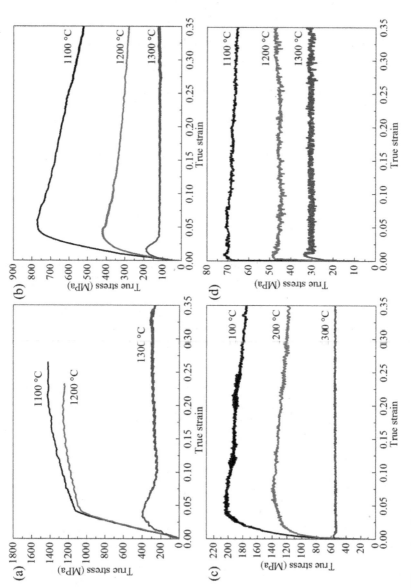

FIGURE 7.1 Variation of true stress with true strain of MoNbTaTiV refractory high entropy alloy at different temperature of deformation and strain rate, (a) 0.5 s^{-1}; (b) 0.05 s^{-1}; (c) 0.005 s^{-1}; (d) 0.0005 s^{-1}. (Reprinted from [38]. Copyright (2021), with permission from Elsevier.)

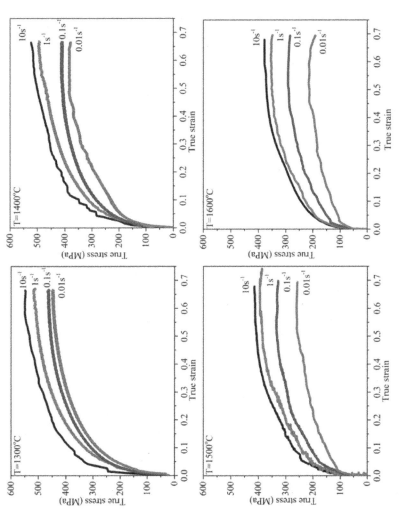

FIGURE 7.2 Variation of true stress with true strain of sintered W-0.25 wt.% Al_2O_3 alloy, studied by simulation-based compression study in hot condition. (Reprinted from [39]. Copyright (2020), with permission from Elsevier.)

of deformation of Mo-1.0%ZrO_2 alloy enhances the plasticity against pure Mo owing to the decrease in the propensity of dislocation and cracks by grain refinement [41]. Development of cavities rather than visible cracks is evident with an increase in the temperature of deformation of Mo-1.0%ZrO_2 as observed in Figure 7.3 (a–e) [41]. The degree of strength enhancement improves with temperature (900–1400°C) and the content of ZrO_2 (from 0.1–1.5 wt.% ZrO_2) [41].

ODS refractory alloys (W-based, Mo-based) are effective in nuclear reactor applications owing to their resistance to embrittlement caused by irradiation. The superior refinement of Mo grains with a larger grain boundary area can capture the defects, reduce the intensity of dislocation loops and cavities, therefore hardening by irradiation is less and resistance to irradiation-based embrittlement is maximum in Mo-ODS alloys compared to low carbon arc-cast (LCAC) Mo, Mo-Ti-Zr (TZM) alloys [42]. The literature also state that high temperature irradiation can cause stress relaxation, redispersal of chemical constituents present at the boundary to improve the plastic flow property and decrease brittleness [42].

7.4 AN APPROACH TO IMPROVE FRACTURE TOUGHNESS

Reducing embrittlement with sufficiently high temperature strength is a prerequisite for high temperature structural applications of refractory alloys. Though oxide dispersion at the grain boundary may negatively impact the ductility, however, appropriate doping of oxides can assist in confining the early failure of the component during applications by impeding crack propagation. Crack deflection or crack bridging mechanisms can improve the fracture toughness in refractory grade ODS alloys. The notches of phases can be changed to a rounded shape to reduce the stress concentration using the double stage heat treatment method to enhance fracture toughness [43]. The enhanced fracture toughness depends on the grain size reduction (higher plastic zone), the reduced size of precipitates, and the plain stress condition for thin samples of ODS Mo alloys [44]. Therefore, the lower the size of the dispersed oxide particles, the higher the grain refinement, so the higher will be the fracture toughness. The DBTT of ODS-Mo in the longitudinal direction is less than -150°C compared to higher DBTT for LCAC Mo (23°C) and TZM alloy (-50°C) [44]. Improving the fracture toughness of W for ballistic application requires reducing the contiguity (W/W interfacial fracture) by incorporating a ductile matrix. The propensity of different fracture types depends on the percentage of brittle constituents. An increase in W content from 90 wt.% to 97 wt.% enhances the crack length by around three times, a decrease in strain in the process region of W increases the fracture between W/W particles, decreases the fracture by the rupture of the ductile phase (NiWFe) (at 23°C) [45]. The paper reports that cessation or blunting of the microcrack through the ductile NiWFe region enhances the fracture toughness [45]. The reduction of the fracture toughness in ODS-Mo alloys is attributed to the coarsening of grains, oxides, and increased porosity, resulting in significant crack formation [46]. The content of the oxides and the processing condition also influence the fracture toughness. Wang et al. show that an increase in Al_2O_3 content up to (0.25 wt.%) increases the fracture toughness both in the sintered and the swaged condition (highest value: 21 MPa$\sqrt{m}$ in swaged alloys), followed by a drop owing

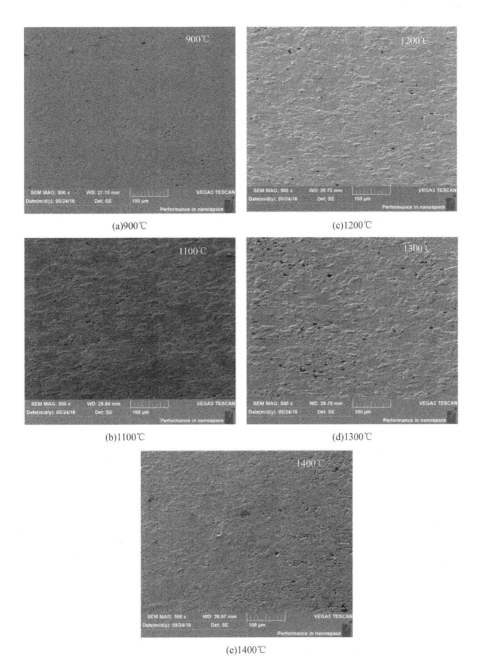

FIGURE 7.3 Deformed microstructure of Mo-1.0%ZrO$_2$ at various deformation temperatures. (Reprinted from [41]. Copyright (2017), with permission from Elsevier.)

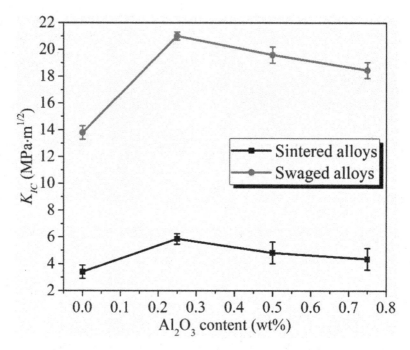

FIGURE 7.4 Variation of fracture toughness (K_{IC}) with Al_2O_3 content in sintered and swaged pure W (0 wt.% Al_2O_3) and W-Al_2O_3 alloys. (Reprinted from [6]. Copyright (2020), with permission from Elsevier.)

to the coarsening of Al_2O_3 leading to lower the hindrance against crack growth (Figure 7.4) [6].

The fracture morphology in Figure 7.5 for sintered pure W, W-Al_2O_3 alloys indicates that the fracture is brittle intergranular, however, with 0.50 wt.% and 0.75 wt.% Al_2O_3 addition it results in a minor transgranular fracture (Figure 7.5 (c), Figure 7.5 (d)), which specifies the improvement in grain boundary bonding [6]. The fracture mode of swaged pure W, W-Al_2O_3 alloys is transgranular (Figure 7.6) and the results suggest that the addition of Al_2O_3 and swaged alloys enriches the fracture toughness of W-Al_2O_3 alloys [6].

The embrittlement generated due to irradiation in W, Mo can be offset by oxide dispersion. The reported fracture toughness values in ODS-Mo alloys (in a longitudinal direction) after irradiation is maximum (26–107 MPa $\sqrt{m}$) compared to LCAC Mo, and TZM alloy is accredited to a thin sheet toughening mechanism, and a reduction in boundary spacing [47]. The presence of O, N at the grain boundary can degrade the fracture toughness, therefore, improving the purity of metal by adding elements such as yttrium (Y), Zr in W can strongly combine O, N to enhance the toughness [48, 49]. Xie et al. have stated that the toughness of WZY alloys can be increased by Zr alloying, hot rolling and dispersion of nanoscale oxide (Y_2O_3) [50]. It is evident from the study that the coarsening issue of nano-oxide is related to increased grain growth and fracture toughness can be reduced by Zr [50]. Cheng et al. have reported

Mechanical Behavior

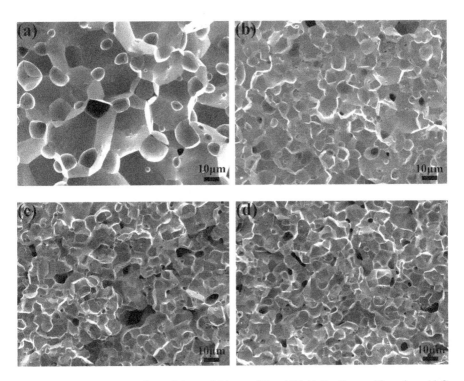

FIGURE 7.5 Fracture surface of sintered (a) pure W and W-Al_2O_3 alloys with various Al_2O_3 addition; (b) 0.25 wt.%; (c) 0.50 wt.%; (d) 0.75 wt.%. (Reprinted from [6].Copyright (2020), with permission from Elsevier.)

the dual effect of a decrease in size of La_2O_3 and a reduction in O concentration of Mo grain boundary by controlled ZrB_2 addition (1.5 wt.%) and this can facilitate an improvement in the percentage elongation with transgranular fracture [51], which may enhance the toughness. Recent reports note that nanostructured BCC NbMoTaW high entropy alloys exhibit higher fracture toughness (2.9–3.3 MPa√m against single crystal BCC NbMoTaW (1.6 MPa√m) and Bi-crystal NbMoTaW alloys (0.2 MPa√m) [52–54]. Interaction between point defects with dislocation in deformed material restricts the dislocation movement and facilitates transgranular fracture by a drop in slip planes available for plastic flow [55]. The formation of intermetallic phases followed by decohesion leads to the development of microvoids and also the presence of the FCC phase in refractory ODS alloys contributes to significant plastic deformation, resulting in ductile fracture [56]. Although improved percentage elongation at elevated temperatures is related to appreciable fracture toughness, tensile testing of powder metallurgy, sintered (1950°C) and cross-rolled processed Mo-La_2O_3-ZrC alloys display higher elongation (10.0 ± 0.7%) at room temperature against 800°C owing to reduced intensity of thin sheet toughening [18], and necking hindrance at 800°C [57]. However, an increase in temperature of the tensile study beyond 1200°C induces the efficient sliding of the grain boundary, the development of dimples assist in the improvement of elongation [57].

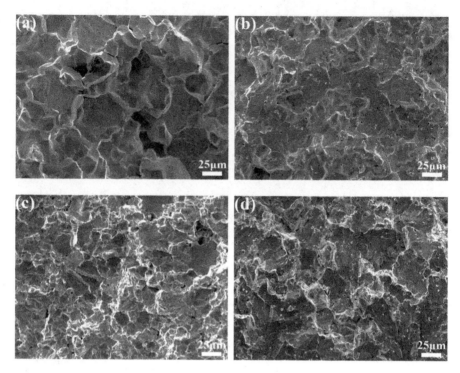

FIGURE 7.6 Fracture surface of swaged (a) pure W and W-Al$_2$O$_3$ alloys with various Al$_2$O$_3$ addition; (b) 0.25 wt.%; (c) 0.50 wt.%; (d) 0.75 wt.%. (Reprinted from [6]. Copyright (2020), with permission from Elsevier.)

7.5 WEAR PROPERTIES

Refractory alloys with high hardness, such as tungsten also possess high wear resistance for applications in space or defense systems [58]. According to the Archard wear model, the hardness and sliding distance, which depend on the sliding time and the applied load, impact the wear property [59]. In this context, hard nanostructured ODS refractory alloys should possess high wear resistance. Apart from hardness, several other factors also influence the wear properties, such as the hardness:elastic modulus, toughness [60], or yield strength:elastic modulus [61]. The temperature of the applications is crucial as brittleness at low temperature and low toughness can counteract the wear resistance [62, 63]. The effect of adding Zr(Y)O$_2$ on the wear behavior of W-Zr(Y)O$_2$ alloys has been studied by Xiao et al. [64]. It is reported that the wear loss decreases with an increase in the percentage mass fraction of the Zr(Y)O$_2$ particles addition up to 3% and a further increase beyond 3 wt.% results in an increase in wear loss [64]. The influence of ZrO$_2$ on the wear study of W-ZrO$_2$ at all applied loads shows an initial drop in wear and minimum wear is attained with 3 wt.% ZrO$_2$ followed by an increase in wear loss (Figure 7.7) [65].

The high hardness of ZrO$_2$ and an increase in strength of W by ZrO$_2$ are responsible for hindering the microcutting process and reduce the wear, though higher

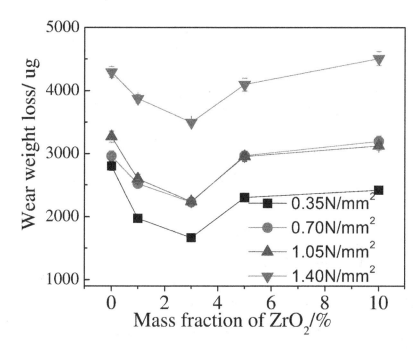

FIGURE 7.7 Variation of wear weight loss with mass fraction of ZrO_2 of W alloy (Reprinted from [65]. Copyright (2017), with permission from Elsevier.)

ZrO_2 addition increases the wear due to the development of phase transformation assisted microcracks, dragging out of ZrO_2 particles at 5.0 wt.% ZrO_2 additions (Figure 7.8) [65].

The increase in hardness and elastic modulus in W-12Co-2Y_2O_3 (wt.%) alloys can also improve wear resistance against W-12Co-2TiC alloys, which is important in coating applications [66]. The literature shows the combined effect of grain size, density, microhardness on the wear resistance of W-ZrO_2 alloys [5]. Though the alloys prepared by the hot swaging process have a lower hardness and coarse grain but the percentage relative density is maximum and the oxide grain size is minimum compared to hydrogen furnace-based sintering and hot isostatic pressing, therefore, the wear loss is minimum for the hot swaging process [5]. Lower porosity can restrict the surface cracking and minimize the wear volume. During continued deformation, the oxide particles can be damaged and can further lower the friction co-efficient [67]. The oxide particles can also provide lubricating to decrease the wear [68]. The resistance to wear also depends on the formation of intermetallic phases in refractory alloys, which improve the strength of the matrix, though the higher volume fraction of intermetallics with less plastic deformation ability also reduces the wear resistance [69]. The literature also provides insightful evidence regarding the lubrication effect, the reduction of cracking and the wear rate by oxides for high entropy alloys ($Mo_{20}Ta_{20}W_{20}Nb_{20}V_{20}$) [70]. The wear rate and temperature relationship at 25°C, 150°C and 300°C for refractory high entropy alloys (HfNbTiZr and

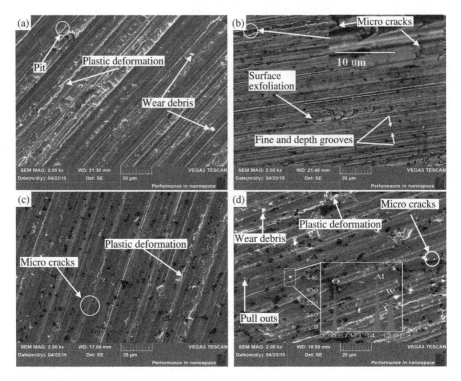

FIGURE 7.8 Wear track images of W-ZrO_2 alloys (a) 0 wt.% ZrO_2; (b) 1.0 wt.% ZrO_2; (c) 3.0 wt.% ZrO_2; (d) 5.0 wt.% ZrO_2. (Reprinted from [65]. Copyright (2017), with permission from Elsevier.)

HfNbTaTiZr) shows that the wear rate increases with the increase in temperature from 25°C to 150°C followed by a reduction at 300°C against 150°C [71]. The abrasive wear at 25°C changes to plowing at 150°C and oxidation wear at 300°C and therefore reduces the wear rate at 300°C compared to 150°C [71]. The success of the oxidation wear to reduce the wear volume depends on the continuity of the oxide film [71] without any blistering or spallation.

7.6 INFERENCES

The above discussion elucidates that adding oxides to refractory metals/alloys improves their hardness and strength. The stress concentration by oxide addition at the grain boundary can cause a drop in the percentage elongation and toughness which can be balanced by suitable tailoring of the microstructure and dispersing the oxides mainly as intragranular particles, retaining the fineness of the dispersed oxides, thus hindering/bending crack growth. Several methods of enhanced grain boundary cohesion by lowering grain boundary impurities (O, N), and restricting the low temperature brittleness have been discussed. Apart from hardness, the percentage densification, the finer oxide grain size, and toughness need to be retained to provide adequate wear resistance of ODS refractory alloys.

REFERENCES

[1] R. Liu, Z.M. Xie, T. Hao, Y. Zhou, X.P. Wang, Q. F. Fang, C. S. Liu, Fabricating high performance tungsten alloys through zirconium micro-alloying and nano-sized yttria dispersion strengthening, *J. Nucl. Mater.*, 451 (1–3) (2014) 35–39.

[2] R. Liu, Z. M. Xie, Q. F. Fang, T. Zhang, X. P. Wang, T. Hao, C. S. Liu, Y. Dai, Nanostructured yttria dispersion-strengthened tungsten synthesized by sol–gel method, *J. Alloys Compd.*, 657 (2016) 73–80.

[3] R. Liu, Y. Zhou, T. Hao, T. Zhang, X.P. Wang, C.S. Liu, Q.F. Fang, Microwave synthesis and properties of fine-grained oxides dispersion strengthened tungsten, *J. Nucl. Mater.*, 424 (1–3) (2012) 171–175.

[4] X.F. Xie, K. Jing, Z.M. Xie, R. Liu, J.F. Yang, Q.F. Fang, C.S. Liu, X. Wu, Mechanical properties and microstructures of W–TiC and W–Y_2O_3 alloys fabricated by hot-pressing sintering, *Mater. Sci. Eng. A*, (2021) 141496.

[5] F. Xiao, Q. Miao, S. Wei, Z. Li, T. Sun, L. Xu, Microstructure and mechanical properties of W-ZrO_2 alloys by different preparation techniques, *J. Alloys Compd.*, 774 (2019) 210–221.

[6] C. Wang, L. Zhang, K. Pan, S. Wei, X. Wu, Q. Li, Effect of Al_2O_3 content and swaging on microstructure and mechanical properties of Al_2O_3/W alloys, *Int. J. Refract. Met. Hard Mater.*, 86 (2020) 105082.

[7] L. Yao, Y. Gao, Y. Huang, Y. Li, X. Huang, P. Xiao, Fabrication of Mo-Y_2O_3 alloys using hydrothermal synthesis and spark plasma sintering, *Int. J. Refract. Met. Hard Mater.*, 98 (2021) 105558.

[8] F. Xiao, Q. Miao, S. Wei, T. Barriere, G. Cheng, S. Zuo, L. Xu, Uniform nanosized oxide particles dispersion strengthened tungsten alloy fabricated involving hydrothermal method and hot isostatic pressing, *J. Alloys Compd.*, 824 (2020) 153894.

[9] S. Coskun, M. L. Ovecoglu, Effects of Y_2O_3 additions on mechanically alloyed and sintered W-4wt.% SiC composites, *Int. J. Refract. Met. Hard Mater.*, 29 (6) (2011) 651–655.

[10] M. A. Yar, S. Wahlberg, H. Bergqvist, H. G. Salem, M. Johnsson, M. Muhammed, Spark plasma sintering of tungsten–yttrium oxide composites from chemically synthesized nanopowders and microstructural characterization, *J. Nucl. Mater.*, 412 (2) (2011) 227–232.

[11] Y. Kim, K. H. Lee, E. P. Kim, D. Ik Cheong, S. H. Hong, Fabrication of high temperature oxides dispersion strengthened tungsten composites by spark plasma sintering process, *Int. J. Refract. Met. Hard Mater.*, 27 (2009) 842–846.

[12] W. M. R. Daoush, A. H. A. Elsayed, O. A. G. E. Kady, M. A. Sayed, O. M. Dawood, Enhancement of physical and mechanical properties of oxide dispersion-strengthened tungsten heavy alloys, *Metall. Mater. Trans. A*, 47 (2016) 2387–2395.

[13] Y. Wang, M. Chen, F. Zhou E. Ma, High tensile ductility in a nanostructured metal, *Nature*, 419 (2002) 912–915.

[14] Y. H. Zhao, J. F. Bingert, X. Z. Liao, B. Z. Cui, K. Han, A. V. Sergueeva, A. K. Mukherjee, R. Z. Valiev, T. G. Langdon, Y. T. Zhu, Simultaneously increasing the ductility and strength of ultra-fine-grained pure copper, *Adv. Mater.*, 18 (22) (2006) 2949–2953.

[15] L. Fan, T. Yang, Y. Zhao, J. Luan, G. Zhou, H. Wang, Z. Jiao, C.-T. Liu, Ultrahigh strength and ductility in newly developed materials with coherent nanolamellar architectures, *Nat. Commun.*, 11 (2020) 6240.

[16] E. Ma, Eight routes to improve the tensile ductility of bulk nanostructured metals and alloys, *JOM*, 58 (2006) 49–53.

[17] G. Liu, G. J. Zhang, F. Jiang, X. D. Ding, Y. J. Sun, J. Sun, E. Ma, Nanostructured high-strength molybdenum alloys with unprecedented tensile ductility, *Nat. Mater.*, 12 (2013) 344–350.

[18] B. V. Cockeram, The mechanical properties and fracture mechanisms of wrought low carbon arc cast (LCAC), molybdenum–0.5pct titanium–0.1pct zirconium (TZM), and oxide dispersion strengthened (ODS) molybdenum flat products, *Mater. Sci. Eng. A*, 418 (1–2) (2006) 120–136.

[19] W. E. Gurwell, in: A. Crowson, E. S. Chen (Eds.), *Tungsten and Tungsten Alloys: Recent Advances*, TMS, Warrendale, PA, (1991) pp. 43–52.

[20] B. H. Rabin, R. M. German, Microstructure effects on tensile properties of tungsten-nickel-iron composites, *Metall. Trans. A*, 19 (1988) 1523–1532.

[21] A. Upadhyaya, Processing strategy for consolidating tungsten heavy alloys for ordnance applications, *Mater. Chem. Phys.*, 67 (1–3) (2001) 101–110.

[22] W. H. Baek, M. H. Hong, S. Lee, D. T. Chung, in: *Tungsten and Refractory Metals*, Vol. 2, *Metal Powder Industries Federation*, Princeton, NJ, (1995) pp. 463–471.

[23] H. J. Ryu, S. H. Hong, W. H. Baek, Microstructure and mechanical properties of mechanically alloyed and solid-state sintered tungsten heavy alloys, *Mater. Sci. Eng. A*, 291 (2000) 91–96.

[24] K. S. Churn, PhD thesis, Korea Advanced Institute of Science and Technology, (1979).

[25] C. Wang, L. Zhang, S. Wei, K. Pan, X. Wu, Q. Li, Effect of ZrO_2 content on microstructure and mechanical properties of W alloys fabricated by spark plasma sintering, *Int. J. Refract. Met. Hard Mater.*, 79 (2019) 79–89.

[26] K. H. Lee, S. I. Cha, H. J. Ryu, S. H. Hong, Effect of two-stage sintering process on microstructure and mechanical properties of ODS tungsten heavy alloy, *Mater. Sci. Eng. A*, 458 (1–2) (2007) 323–329.

[27] Z. M. Xie, et al., Extraordinary high ductility/strength of the interface designed bulk W-ZrC alloy plate at relatively low temperature, *Sci. Rep.*, 5 (2015) 16014.

[28] J. Reiser, J. Hoffmann, U. Jantsch, M. Kilmenkov, S. Bonk, C. Bonnekoh, M. Rieth, A. Hoffmann, T. Mrotzek, Ductilisation of tungsten (W): On the shift of the brittle-to-ductile transition (BDT) to lower temperatures through cold rolling, *Int. J. Refract. Met. Hard Mater.*, 54 (2016) 351–369.

[29] M. Mabuchi, K. Okamoto, N. Saito, T. Asahina, T. Igarashi, Deformation behavior and strengthening mechanisms at intermediate temperatures in W-La_2O_3, *Mater. Sci. Eng. A*, 237 (1997) 241–249.

[30] M. V. Aguirre, A. Martín, J. Y. Pastor, J. Llorca, M. A. Monge, R. Pareja, Mechanical behavior of W-Y_2O_3 and W-Ti alloys from 25 °C to 1000 °C, Metall. *Mater. Trans. A*, 40 (10) (2009) 2283–2290.

[31] A. Luo, D. L. Jacobson, K. S. Shin, Solution softening mechanism of iridium and rhenium in tungsten at room temperature, *Int. J. Refract. Met. Hard Mater.*, 10 (2) (1991) 107–114.

[32] C. Ren, Z. Z. Fang, M. Koopman, B. Butler, J. Paramore, S. Middlemas, Methods for improving ductility of tungsten: A review, *Int. J. Refract. Met. Hard Mater.*, 75 (2018) 170–183.

[33] R. Bianco, R. W. Buckman Jr., Mechanical properties of oxide dispersion strengthened (ODS) molybdenum alloys, WAPD-T-3175, Bettis Atomic Power Lab., West Mifflin, PA, (1998).

[34] H. K. D. H. Bhadeshia, Recrystallisation of practical mechanically alloyed iron-base and nickel-base superalloys, *Mater. Sci. Eng. A*, 223 (1–2) (1997) 64–77.

[35] K. Mino, Hiroshi Harada, H. K. D. H. Bhadeshia, M. Yamazaki, Mechanically alloyed ODS Steel & Ni-base superalloy: A study of directional recrystallisation using DSC, *Mater. Sci. Forum*, 88–90 (1992) 213–220.

[36] S. Park, D. K. Kim, S. Lee, H. J. Ryu, S. H. Hong, H. J. Ryu, Dynamic deformation behavior of an oxide-dispersed tungsten heavy alloy fabricated by mechanical alloying, *Metall. Mater. Trans. A*, 32 (2001) 2011–2020.

[37] C. M. Sellars, W. J. McTegart, On the mechanism of hot deformation, *Acta Metall.*, 14 (9) (1966) 1136–1138.

[38] Q. Liu, G. Wang, Y. Liu, X. Sui, Y. Chen, S. Luo, Hot deformation behaviors of an ultrafine-grained MoNbTaTiV refractory high-entropy alloy fabricated by powder metallurgy, *Mater. Sci. Eng. A*, 809 (2021) 140922.

[39] C. Wang, L. Zhang, S. Wei, X. Li, X. Wu, Q. Li, K. Pan, Establishment of processing map, microstructure and high-temperature tensile properties of W-0.25 wt.% Al_2O_3 alloys, *J. Alloys Compd.*, 831 (2020) 154751.

[40] Y. V. R. K. Prasad, H. L. Gegel, S. M. Doraivelu, J. C. Malas, J. T. Morgan, K. A. Lark, D. R. Barker, Modeling of dynamic material behavior in hot deformation: Forging of Ti-6242, *Metall. Mater. Trans. A*, 15 (1984) 1883–1892.

[41] C. Cui, Y. Gao, S. Wei, G. Zhang, Y. Zhou, X. Zhu, Microstructure and high temperature deformation behavior of the Mo-ZrO_2 alloys, *J. Alloys Compd.*, 716 (2017) 321–329.

[42] T. S. Byun, M. Li, B. V. Cockeram, L. L. Snead, Deformation and fracture properties in neutron irradiated pure Mo and Mo alloys, *J. Nucl. Mater.*, 376 (2) (2008) 240–246.

[43] C. Ye, H. Li, L. Jia, Z. Jin, G. Sun, H. Zhang, Improvement of fracture toughness of Nb-Si alloy by two-step heat treatment, *Int. J. Refract. Met. Hard Mater.*, 94 (2021) 105378.

[44] B. V. Cockeram, The role of stress state on the fracture toughness and toughening mechanisms of wrought molybdenum and molybdenum alloys, *Mater. Sci. Eng. A*, 528 (1) (2010) 288–308.

[45] M. E. Alam, G. R. Odette, On the remarkable fracture toughness of 90 to 97W-NiFe alloys revealing powerful new ductile phase toughening mechanisms, *Acta Mater.*, 186 (2020) 324–340.

[46] B. V. Cockeram, The fracture toughness and toughening mechanism of commercially available unalloyed molybdenum and oxide dispersion strengthened molybdenum with an equiaxed, large grain structure, *Metall. Mater. Trans. A*, 40 (2009) 2843.

[47] B. V. Cockeram, T. S. Byun, K. J. Leonard, J. L. Hollenbeck, L. L. Snead, Post-irradiation fracture toughness of unalloyed molybdenum, ODS molybdenum, and TZM molybdenum following irradiation at 244 °C to 507 °C, *J. Nucl. Mater.*, 440 (1–3) (2013) 382–413.

[48] Y. Ueda, et al., Recent progress of tungsten R&D for fusion application in Japan, *Phys. Scr.*, T145 (2011) 014029.

[49] S. Wurster, et al., Recent progress in R&D on tungsten alloys for divertor structural and plasma facing materials, *J. Nucl. Mater.*, 442 (2013) S181–S189.

[50] Z. M. Xie, R. Liu, T. Zhang, Q. F. Fang, C. S. Liu, X. Liu, G. N. Luo, Achieving high strength/ductility in bulk W-Zr-Y_2O_3 alloy plate with hybrid microstructure, *Mater. Des.*, 107 (2016) 144–152.

[51] P. M. Cheng, S. L. Li, G. J. Zhang, J. Y. Zhang, G. Liu, J. Sun, Ductilizing Mo–La_2O_3 alloys with ZrB_2 addition, *Mater. Sci. Eng. A*, 619 (2014) 345–353.

[52] O. N. Senkov, G. B. Wilks, J. M. Scott, D. B. Miracle, Mechanical properties of $Nb_{25}Mo_{25}Ta_{25}W_{25}$ and $V_{20}Nb_{20}Mo_{20}Ta_{20}W_{20}$ refractory high entropy alloys, *Intermetallics*, 19 (2011) 698–706.

[53] Y. Zou, P. Okle, H. Yu, T. Sumigawa, T. Kitamura, S. Maiti, W. Steurer, R. Spolenak, Fracture properties of a refractory high-entropy alloy: In situ micro-cantilever and atom probe tomography studies, *Scripta Mater.*, 128 (2017) 95–99.

[54] Y. Xiao, Y. Zou, H. Ma, A. S. Sologubenko, X. Maeder, R. Spolenak, J. M. Wheeler, Nanostructured NbMoTaW high entropy alloy thin films: High strength and enhanced fracture toughness, *Scr. Mater.*, 168 (2019) 51–55.

[55] M. A. Zikry, M. Kao, Inelastic microstructural failure mechanisms in crystalline materials with high angle grain boundaries, *J. Mech. Phys. Solids*, 44 (1996) 1765–1798.

[56] W. Li, D. Xie, D. Li, Y. Zhang, Y. Gao, P. K. Liaw, Mechanical behavior of high-entropy alloys, *Prog. Mater. Sci.*, 118 (2021) 100777.

[57] J. Gan, Q. Gong, Y. Jiang, H. Chen, Y. Huang, K. Du, Y. Li, M. Zhao, F. Lin, D. Zhuang, Microstructure and high-temperature mechanical properties of second-phase enhanced Mo-La_2O_3-ZrC alloys post-treated by cross rolling, *J. Alloys Compd.*, 796 (2019) 167–175.

[58] H. Bolt, V. Barabash, W. Krauss, J. Linke, R. Neu, S. Suzuki, N. Yoshida, A.U. Team, Materials for the plasma-facing components of fusion reactors, *J. Nucl. Mater.*, 329–333 (2004) 66–73.

[59] J. F. Archard, Contact and rubbing of flat surfaces, *J. Appl. Phys.*, 24 (1953) 981–988.

[60] A. Leyland, A. Matthews, On the significance of the H/E ratio in wear control: A nanocomposite coating approach to optimised tribological behaviour, *Wear*, 246 (2000) 1–11.

[61] T. W. Liskiewicz, B. D. Beake, N. Schwarzer, M. I. Davies, Short note on improved integration of mechanical testing in predictive wear models, *Surf. Coating Technol.*, 237 (2013) 212–218.

[62] D. Q. Ma, J. P. Xie, J. W. Li, S. Liu, F. M. Wang, H. J. Zhang, Synthesis and hydrogen reduction of nano-sized copper tungstate powders produced by a hydrothermal method, *Int. J. Refract. Met. Hard Mater.*, 46 (2014) 152–158.

[63] S. V. Madge, D. V. L. Luzgin, J. J. Lewandowski, A. L. Greer, Toughness, extrinsic effects and Poisson's ratio of bulk metallic glasses, *Acta Mater.*, 60 (2012) 4800–4809.

[64] F. Xiao, Q. Miao, T. Barriere, G. Cheng, S. Zuo, L. Xu, A study on the effect of solution acidity on the microstructure, mechanical, and wear properties of tungsten alloys reinforced by yttria-stabilised zirconia particles, *Mater. Today Commun.*, 27 (2021) 102223.

[65] F. Xiao, L. Xu, Y. Zhou, K. Pan, J. Li, W. Liu, S. Wei, Preparation, microstructure, and properties of tungsten alloys reinforced by ZrO_2 particles, *Int. J. Refract. Met. Hard Mater.*, 64 (2017) 40–46.

[66] C. L. Chen, Sutrisna, Study of W-Co ODS coating on stainless steels by mechanical alloying, *Surf. Coat. Technol.*, 350 (2018) 954–961.

[67] R. Yamanoglu, E. Karakulak, A. Zeren, M. Zeren, Effect of heat treatment on the tribological properties of Al–Cu–Mg/nanoSiC composites, *Mater. Des.*, 49 (2013) 820–825.

[68] G. Cui, J. Han, G. Wu, High-temperature wear behavior of self-lubricating Co matrix alloys prepared by P/M, *Wear*, 346–347 (2016) 116–123.

[69] L. Wang, D. Y. Li, Effects of yttrium on microstructure, mechanical properties and high-temperature wear behavior of cast Stellite 6 alloy, *Wear*, 255 (2003) 535–544.

[70] A. Poulia, E. Georgatis, A. Lekatou, A. E. Karantzalis, Microstructure and wear behavior of a refractory high entropy alloy, *Int. J. Refract. Met. Hard Mater.*, 57 (2016) 50–63.

[71] M. Sadeghilaridjani, M. Pole, S. Jha, S. Muskeri, N. Ghodki, S. Mukherjee, Deformation and tribological behavior of ductile refractory high-entropy alloys, *Wear*, 478–479 (2021) 203916.

8 High Temperature Oxidation Characteristics

8.1 OXIDATION OF REFRACTORY ALLOYS

The oxidation kinetics for refractory metals depend on temperature and atmospheric pressure. The oxidation proceeds with a minimal degree at temperature < 200°C and the oxidation rate law changes to parabolic between 300–400°C [1]. The tendency for oxide formation also depends on the oxygen solubility in refractory metals such as W. Mo exhibits low oxygen solubility, therefore oxide scale formation occurs at the reduced pressure of oxygen [1]. The oxidation resistance is accredited to minimize porosities and the diffusion channel for oxygen or volatile constituents. Continuous scale formation is not the only sufficient condition for oxidation resistance. Oxidation stresses generated due to new oxide phase evolution and thermal stress (heating-cooling cycle) also impact the extent of the oxide scale bonding with the substrate or blistering/peeling off. The density of the oxide scale, the oxide grain size, the morphology, the molar volume, the thermal expansion coefficient of the oxides also influence the passive layer formation. The stability of the oxides is also affected by the vapor pressure and at an elevated temperature, high vapor pressure of oxides compared to atmospheric pressure leads to a reduction in the oxidation resistance. An oxidation study of NbTaTi, NbTaZr, C-3009(Nb-20Hf-6 W, at%), C-103 (Nb-5.4Hf-2.0Ti-0.7Zr-0.3Ta-0.3 W, at%) shows the normalized mass gain increases with an increase in oxidation time for all the alloys, though C-3009 alloy shows lower mass gain at a higher time against the other alloys owing to the presence of Hf rather than Ta, which forms a less effective oxide layer (Ta_2O_5) to protect the substrate [2–4]. Nb-based alloys used in rocket engines in the temperature regime of 600°C–800°C exhibit significant oxidation and affect the durability of the rocket engines [5, 6]. An oxidation study of Hf-27Ta, Hf-21Mo-21Ta, Hf-21Nb-21Ta and Hf-21Ta-21W (in at%) at 1600°C shows that oxide sublimation and melting seriously impact the oxidation resistance [7]. Maximum weight gain has been reported around 230 mg/cm^2/h at 1 h of oxidation for Hf-21Nb-21Ta and Hf-27Ta is the highest oxidation-resistant alloy [7]. The mass gain per surface area during oxidation of Nb-Cr-V-W-Ta high entropy alloy increases with an increase in the oxidation temperature and time, however, a parabolic trend is observed [8]. The paper illustrates that Nb-Cr-V-W-Ta high entropy alloy shows inadequate oxidation resistance in the air beyond 1000°C [8]. The activation energy of oxygen diffusion in W is 80 kJ/mol compared to oxygen diffusion in Nb (107 kJ/mol) [9], but the formed WO_3 oxides in W may sublime at temperatures > 1000°C, which significantly deteriorates the high-temperature applicability (higher than 1000°C) of W [10]. Li et al. have reported that an increase in W content from 10 wt.% to 30 wt.% in Nb-W alloys reduces the parabolic rate constant from 6640.62 mg^2 cm^{-4} h^{-1} to 5884.42 mg^2 cm^{-4} h^{-1} during oxidation at 1300°C for

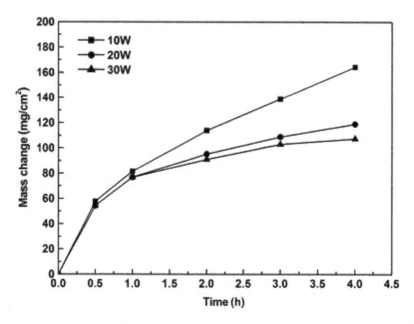

FIGURE 8.1 Cyclic oxidation plot for Nb-10W, Nb-20W, Nb-30W alloys at 1300°C. (Reprinted from [11]. Copyright (2012), with permission from Elsevier.)

4 h [11]. It is also evident from the paper that until 0.5 h of oxidation all the alloys show an almost similar linear rise in oxidation kinetics, however, beyond that, the mass change per unit area shows the minimum change for Nb 30W compared to Nb 10W, Nb-20W (Figure 8.1) [11].

A reduction in W content and the proper addition of Cr can inhibit the formation of volatile WO_3 and improve the oxidation resistance in W-Cr-Nb alloys [12]. $(W_{0.5}Cr_{0.5})_{90}Nb_{10}$ and $(W_{0.4}Cr_{0.6})_{90}Nb_{10}$ alloys display enriched more percentage oxidation resistance than $(W_{0.7}Cr_{0.3})_{90}Nb_{10}$, owing to the formation of Cr_2WO_6 and the $Nb_2O_5 \cdot 3WO_3$ oxide scale [12]. This is supported by the lower grain size of the W matrix in $(W_{0.5}Cr_{0.5})_{90}Nb_{10}$ and $(W_{0.4}Cr_{0.6})_{90}Nb_{10}$ alloys to increase the Cr diffusion through higher availability of the grain boundaries for Cr oxide formation and superior oxidation resistance [12]. The poor oxidation resistance of Mo at elevated temperature can be counteracted by the combined addition of Si and B. The evaporation of B_2O_3 increases with an increase in temperature and enriches the viscosity of the borosilicate layer which further reduces the oxidation by hindering the oxygen diffusion [13]. Reduction in cracking during oxidation also may direct enhancement in the resistance of oxidation. The addition of Ta in W results in 30.7% decrease in crack density by decreasing the grain size compared to pure W [14]. Applying a coating of Mo-Si-B to an Mo-W-Si-B alloy is advantageous to prevent oxidation [15]. Without a coating, oxidation resistance improved at 1300°C compared to 1100°C owing to the faster development of the borosilicate layer and the formation of the lower oxides of W (WO_2) at a higher temperature [15]. Experimental evidence shows that the oxidation resistance of W-Cr-Zr alloys is superior compared to W-Cr-Y alloy and the extent of resistance depends on Zr: Cr and the ratio needs to be optimized to improve the Cr_2O_3 passive layer formation and to enhance the bonding between the oxide scale and the substrate [16]. The mass change of W, W-Re alloys in a fusion reactor application depends on the irradiation flux of He^+, and 0.3–1.0 wt.% addition of Re in pure W increases the mass loss and erosion rate, however, it decreases with 3.0 wt.% Re addition, though the drop is marginal compared to pure W [17]. The effect of increased Cr addition and temperature on oxidation of Nb-W-Cr alloys shows a lower weight gain at a higher temperature (1300°C) oxidation compared to 900°C and the trend is evident for high Cr containing alloys (25 wt.%, 30 wt.%) [18]. The oxidation rate constant with 25 wt.% and 30 wt.% Cr addition is 333.5 $mg^2\ cm^{-4}\ h^{-1}$, 308.9 $mg^2\ cm^{-4}\ h^{-1}$ respectively at 900°C, which reduces to 108.1 $mg^2\ cm^{-4}\ h^{-1}$, 65.3 $mg^2\ cm^{-4}\ h^{-1}$ at 1300°C with identical composition respectively [18]. An enhancement of resistance against oxidation with an increase in temperature for W-Cr-Pd and Mo-W-Cr-Pd alloys has been reported by Lee and Simkovich, although Mo-W-Cr-Pd alloys have superior resistance during isothermal oxidation than cyclic oxidation as well as against W-Cr-Pd alloys [19]. The oxidation mechanism of W-Cr-Pd alloys presented in the literature suggests four different stages (Figure 8.2): (a) initial development of WO_3, Cr_2WO_6, Cr_2O_3 oxides and temperature-dependent evaporation of WO_3; (b) enriched formation of Cr_2O_3 oxides at subsurface layer; (c) formation of Cr depleted region under the Cr_2O_3 and the depleted zone contains Pd, W, along with Cr diffusion in the Pd-rich region and the presence of Kirkendall voids; and (d) the final stage is related to the increased thickness of the depleted region of Cr [19].

8.2 EFFECT OF OXIDE DISPERSION ON THE OXIDATION OF REFRACTORY ALLOYS

Section 8.1 discusses the factors that influence the oxidation of refractory alloys. Though several investigations suggest enhancing oxidation resistance of refractory metals by alloys addition [14, 15, 16, 18, 19], several studies also indicate the effect of oxide dispersion on the oxidation behavior of refractory metals and alloys. Improvements in densification by oxide dispersion can suppress the oxygen diffusion and volatilization of oxides. Addition of Y_2O_3 to improve the oxidation resistance of W is reported in papers [20, 21]. Y_2O_3 addition in W can facilitate crack deflection and crack growth restriction and further enhance the oxidation resistance [22]. The extent of mass loss and erosion rate of W alloys by He^+ irradiations depends on the content of dispersed oxides [17]. The mass loss and erosion rate marginally drop with the increase in La_2O_3 dispersion from 0.1 vol.% to 1.0 vol.% in W, though a further increase in vol.% of La_2O_3 to 5% significantly enhances both mass loss and the erosion rate and Y_2O_3 dispersion produces a similar trend [17]. It is reported in the study that La_2O_3 is more effective in reducing the erosion rate compared to Y_2O_3 [17]. The mass loss and erosion rate by the combined addition of Re and 1.0 vol%La_2O_3 are higher compared to the addition of only 1.0 vol%La_2O_3 in W (Figure 8.3), and the thickness of the fuzz cover of W reduces in La_2O_3, Y_2O_3 dispersed W alloy against pure W [17].

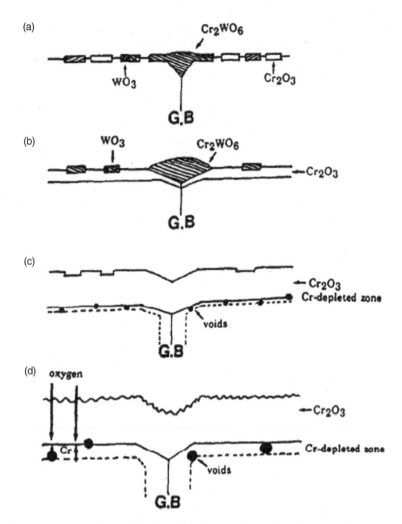

FIGURE 8.2 Stages during oxidation of W-Cr-Pd alloys, (a) initial oxidation; (b) selective oxidation; (c) generation of chromium-depleted zone; (d) after (c). (Reprinted from [19]. Copyright (1990), with permission from Elsevier.)

Another report shows the attainment of appreciable oxidation resistance of Y_2O_3, Al_2O_3 dispersed W-Ni alloys at 1000°C with 10 h of oxidation time against La_2O_3 dispersion and the degree of spallation depends on the sintering temperature and densification [23]. The sublimation rate of WO_3, MoO_3 oxides can be reduced by the formation of more stable oxides by reaction with the alloying elements of dispersed oxides. The dispersed oxide should form a continuous oxide scale rather than segregate and, therefore, the selection of wt.% of oxides is of immense importance. Refinement of grains by oxide dispersion can enhance the plasticity of the oxide

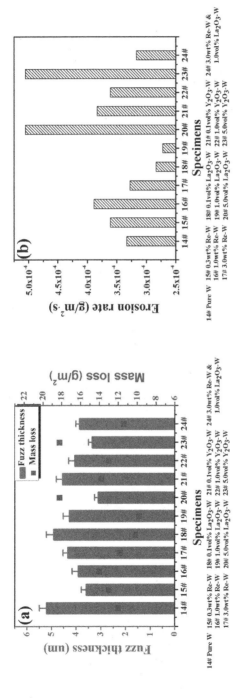

FIGURE 8.3 Variation of (a) thickness of fuzz and mass loss, (b) erosion rate for different W specimens [He+ irradiations temperature: 1420 K, He+ energy: 100 eV, He+ fluence: $5.83 \times 10^{26}/m^2$]. (Reprinted from [17]. Copyright (2019), with permission from Elsevier.)

film and enrich the bonding of the oxide layer [24]. The Gibbs free energy of WO_3, Cr_2O_3, Y_2O_3 at 1273 K is -3.6×10^2 kJ/ (mol O_2), -5.3×10^2 kJ/(mol O_2), -1.4×10^3 kJ/ (mol O_2), respectively and therefore, Y_2O_3 is very stable compared to WO_3, Cr_2O_3 at 1000°C and can hinder oxidation [25]. The ionic radius also contributes to the extent of diffusion of species [26]. Jehanno et al. have reported that powder metallurgically fabricated Y_2O_3 and La_2O_3 dispersed Mo-Si-B alloy exhibit a weight increase at a lower oxidation temperature (650°C) and drop in weight at a higher temperatures (1100°C and 1315°C) [27]. However, the extent of the weight increase or decrease is less in La_2O_3 dispersed alloy at all temperatures and it is summarized that the dispersed oxide reduces the oxygen entrance at the grain boundary and crack confinement, which improves the oxidation resistance [27]. During oxidation of ODS refractory alloys, the major aspect of oxides is promoting the rapid impervious oxide scale and microstructural uniformity [27]. The oxidation resistance of ODS alloys depends on the oxide scale growth mechanism and scale thickness, which may differ in refractory alloys. Sublimation of oxides with high vapor pressure leads to the development of pores in the oxide scales.

The difference is between the thermal expansion coefficient between the formed oxides and the matrix and, with the growth of oxide scales, stresses develops. The deformation ability of the oxide scale is another prime factor which is related to the degree of stress development in the oxides. The literature indicates that the presence of Y_2O_3, $Y_3Al_5O_{12}$ at the grain boundary can restrict the deformation of α-Al_2O_3 during growth [28]. Other studies indicate that elemental Y is effective in enhancing the deformability of α-Al_2O_3 and stress minimization [29–31]. The prevention of vacancy development by oxide dispersion [31] and refinement of the oxide scale [32] can improve the oxide scale behavior and overall oxidation resistance. Impeding the short circuiting channel for the diffusion of cations and anions by oxides will reduce the oxide layer thickness and the stress [33], which may further minimize oxidation. The coefficient of the thermal expansion (CTE) of the formed oxides depends on the oxidation temperature and it is reported that $Y_2W_3O_{12}$ is a product formed during oxidation of W-Cr-Y_2O_3 at a high temperature (1200°C) [20] and has a negative CTE of -7.5 ppm/K within the studied temperature range of 200–800°C [34]. Oxides such as Y_2WO_6, Y_6WO_{12} exhibit positive CTE within 200–800°C, though are stable beyond 1050°C [35]. The lowering of the CTE value to stabilize the oxide microstructure may enhance the resistance to oxidation and also is impacted by the proportionate weight ratio of the oxides [34]. The location of the oxide scale formation depends on the partial pressure of the oxygen (pO_2) which can be evaluated from the Gibbs free energy of formation of oxides, and oxides with a lower partial pressure indicate the formation at the subsurface area with a lower base metal:oxygen ratio. It is reported that the presence of Nb_2O_5. TiO_2 or Nb_2O_5 provides inadequate protection compared to the formation of mixed oxide of $CrNbO_4$ against oxidation [36]. The addition of La in TZM alloys is effective in reducing the mass loss during oxidation for 10, 15, 30 min [37]. The investigation manifests that until oxidation from 300–600°C, the improvement is non-significant and beyond 600–1000°C, the oxidation resistance becomes considerable against TZM alloy which is attributed to the dense grain boundary to minimize the oxygen diffusion [37].

High Temperature Oxidation Characteristics   155

8.3 FACTORS INFLUENCING THE OXIDATION OF OXIDE DISPERSION STRENGTHENED ALLOYS

The oxidation kinetics law of metals or alloys is associated with weight change (Δm), unit surface area (S), oxidation time (t), rate constant (k) and oxidation exponent (n) and the relation is as follows [38]:

$$\left(\frac{\Delta m}{s}\right)^n = kt \qquad (8.1)$$

Depending on the n value, the oxidation law can be linear (n = 1), parabolic (n = 2), or cubic (n = 3). The parabolic law as represented in Equation (8.2) describes how the oxidation rate is lower compared to the linear law:

$$\left(\frac{\Delta m}{s}\right)^2 = k_p t + C \qquad (8.2)$$

where k_p is the parabolic rate constant, C is the constant.

Porous alloys exhibit a higher oxidation rate (linear law) unless the scale-metal adhesion is improved and an impervious scale forms, which shifts the kinetics to the parabolic rate.

During the oxidation of alloys, the oxidation temperature, time, the oxygen partial pressure, the melting point, the sublimation temperature of the oxides, the oxide grain size, the alloy density, the molar volume of oxide, and the scale thickness all influence the oxidation behavior and decide the sustainability of the alloys. The more alloying elements are present, the higher the compounds in the oxides formation and the more complex will be the oxidation mechanism. Specifically for ODS alloys, the fine grain size and crack-arresting mechanism by the dispersed oxides may change the oxide scale growth kinetics and the oxygen diffusion through the grain boundary. The solubility of the oxygen in refractory metals illustrates the degree of oxide formation when exposed to a high temperature. The oxygen solubility at 1000°C in W, Nb, Ta, Mo is 0.03 at.% [39], 2.5 at.% [40], 3 at.% [41], and 0.03 at.% [39], respectively. Therefore, W, Mo will be more likely to be consumed by oxygen to form oxides rather than Nb, Ta. The partial pressure of oxygen at 1000°C for WO_3 and Y_2WO_6 is 4.63×10^{-15} atm [38] and 3.80 ×10^{-28} atm, respectively [23]. It reflects the formation of more stable Y_2WO_6 (reaction product of WO_3 and Y_2O_3) and occurs under the oxide layer of WO_3 through migration of Y^{+3} to the WO_3 layer. The Pilling–Bedworth ratio (PBR) is an important parameter to understand the extent of the compressive stress growth in the oxide scale. Oxides of W, Nb, Ta, such as WO_3, Nb_2O_5, Ta_2O_5, have a PBR of 3.3 [42], 2.68 [43], and 2.5 [44], respectively. Adding oxides such as ZrO_2, Cr_2O_3, Al_2O_3, TiO_2 is associated with a comparatively lower PBR of 1.56, 2.07, 1.28, 1.78, respectively [43]. It is evident from the research findings that ΔG° for WO_3, MoO_3 is comparatively higher against Cr_2O_3, TiO_2, Al_2O_3, ZrO_2 within the studied temperature of 400–1200° C [45]. The difference of ΔG° between refractory oxides of Nb_2O_5, Ta_2O_5 with TiO_2, Cr_2O_3 is not considerable compared to highly stable Al_2O_3, ZrO_2 [45]. During the oxidation of refractory alloys, formation of volatile oxides may leave residual porosity during evaporation and further increase the cationic and oxygen diffusion. Elements with higher affinity for oxygen can form oxides at the surface, the free energy of formation decides the oxide's stability. The WO_3, MoO_3 diffusion results in decay of the base metal and alloy. Therefore, the presence of Y_2O_3 minimizes the diffusion of cation to reduce the oxide volatilization [33]. Though WO_3, MoO_3 (high oxygen: metal ratio) formation is disadvantageous to protect the substrate, sufficient oxygen activity to form stable oxides is necessary, such as SiO_2 (oxygen: metal = 2:1) rather than volatile SiO (oxygen: metal = 1:1) [46]. The cationic radius of added oxides affects the movement of cations. It is reported that ions with a higher radius impede the movement of cations through the grain boundary and facilitate oxygen diffusion [47]. The higher ionic radius of La against other rare-earth ions (Pr, Gd, Sm, Nd, Ce, Y, Er, Dy, Yb) improves the oxidation resistance in Fe-Cr alloys [48]. The paper also investigates how a needle, rod, or cube-shaped oxide morphology can possess a higher oxidation resistance than platelet/cluster-shaped oxides [48]. The above research explains the influence of microstructure and cationic radius of dispersed oxide in refractory metals/alloys on oxidation resistance. The stress development in oxides (σ_m) is related to Young's modulus of oxide (E_{ox}) and metal (E_m), the thermal expansion coefficient of oxide (α_{ox}) and the metal (α_m), thickness of oxide (ξ), ½ (thickness of specimen) (h), temperature difference ($\Delta T = T_{ox}-T$), where T_{ox} = oxidation temperature (T_{ox}), T: temperature at which the stress is evaluated and the Poisson ratio (when the Poisson ratio of oxide and metal is identical) (v) [49]:

$$\sigma_1 = -\frac{E_{ox}\Delta T(\alpha_m - \alpha_{ox})}{(1-v)\left(1+\frac{E_{ox}}{E_m}\frac{\xi}{h}\right)} \quad (8.3)$$

The negative sign indicates the compressive stress on the oxide layer and if the thickness of the oxide scale is less compared to metal, then Equation (8.1) can be presented as:

$$\sigma_1 = -\frac{E_{ox}\Delta T(\alpha_m - \alpha_{ox})}{(1-v)} \quad (8.4)$$

As, $\alpha_m > \alpha_{ox}$ and the stresses on the oxide layer are proportional to the difference in the thermal expansion coefficient between the metal and the oxide, therefore, the formation of oxides that have a thermal expansion coefficient closer to metal will lower the stress and prevent oxide spallation. The crack growth in the oxide layer also depends on the critical stress intensity factor (K_{1c}) and K_{1c} depends on E_{ox} and the surface energy (γ_s) and rate of critical energy release (G_{1c}) as [50, 51]:

$$\frac{K_{1c}^2}{E_{ox}} = G_{1c} = 2\gamma_s \quad (8.5)$$

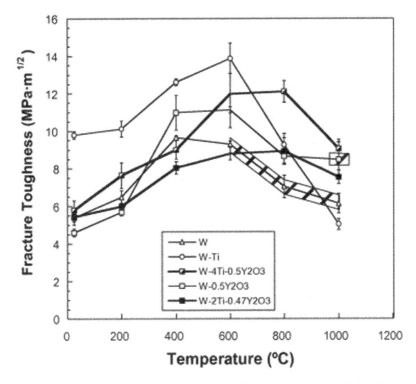

FIGURE 8.4 Variation of fracture toughness with test temperature, (value in shaded region: "apparent fracture toughness," the experiment includes considerable plastic deformation prior to maximum load. (Reprinted from [52], Copyright (2011), with permission from Elsevier.)

The reported K_{1c} for Al_2O_3, Cr_2O_3, NiO is 2.5 MN m$^{-3/2}$, 1.8 MN m$^{-3/2}$, 1.3 MN m$^{-3/2}$, respectively [50] which describes how the stress requirement for cracking is higher in Al_2O_3 than Cr_2O_3, NiO. The effect of oxide dispersion (Y_2O_3) on oxidation of W and W-Ti alloys has been described by high temperature fracture toughness measurements [52]. Figure 8.4 shows how the addition of 0.5 wt.% Y_2O_3 in W, 0.47 wt.%/0.5 wt.% Y_2O_3 in W-Ti alloys improves the fracture toughness of the alloys at a higher temperature (1000°C), however, 0.5 wt.% Y_2O_3 has a superior impact on the enhancement of fracture toughness of W-Ti alloys and therefore contributes to the enrichment of the oxidation resistance [52].

8.4 METHODS TO ENHANCE OXIDATION RESISTANCE

Several factors influencing the oxidation resistance of ODS refractory alloys have been discussed in previous sections (Sections 8.1, 8.2, 8.3). The rapid formation of impervious oxide scale reduces the chance of the cationic transport of base metal and it is specifically essential for oxides with higher vapor pressure than atmospheric pressure. In ODS refractory alloys, several mixed oxides can also form rather than dispersed Y_2O_3, such as Cr_2WO_6, $Y_2W_3O_{12}$ in W-Cr-Y_2O_3 alloys at higher oxidation temperatures (1000°C, 1200°C) [20]. The formation of newer oxides can trigger the variation of the thermal expansion coefficient and result in the spallation of oxides unless the melting of the mixed oxide ($Y_2W_3O_{12}$) leads to viscoplastic flow and pore filling by the liquid phase to improve the oxidation resistance of the W-Cr-Y_2O_3 alloy [20]. The refinement of grains can improve the plasticity of the oxide scale, and uniform distribution of dispersed oxides is responsible for restricting the oxide scale failure [24]. Therefore, the uniform dispersion of oxides is of significant importance in this context. The purity of the alloys (absence of S, C, P) also can contribute to steady oxide layer formation and enrich the oxidation resistance [42]. The pack cementation process involving the formation of intermetallic phases along with the presence of an oxide phase can support the improvement of the oxidation resistance of refractory alloys. Researchers have employed aluminum (88%) and silicon (12%), halide (NH_4Cl), and Al_2O_3 as filling agents on Mo-Si-B for the pack cementation process [53]. The literature indicates that pack cementation reduces the mass loss in Mo-Si-B alloys compared to uncoated Mo-Si-B alloys (Figure 8.5) [53]. The process develops a layered structure as displayed in Figure 8.6 [53]. Other reports suggest that the formation of the intermetallic phase leads to cracking due to the CTE difference between the matrix and the intermetallic phase [54]. Ulrich and Galetz suggest that

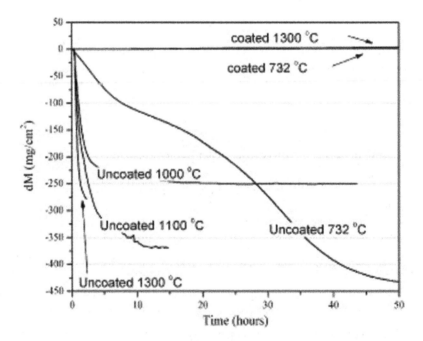

FIGURE 8.5 Oxidation kinetics of pack aluminized and uncoated samples at 50 h of oxidation between 732°C and 1300°C. The plot has been interrupted after oxidation of uncoated samples at 1300°C for 2.5 h owing to considerable condensation of MoO_3 in a thermo-gravimetric analyzer. (Reprinted from [53]. Copyright (2006), with permission from Elsevier.)

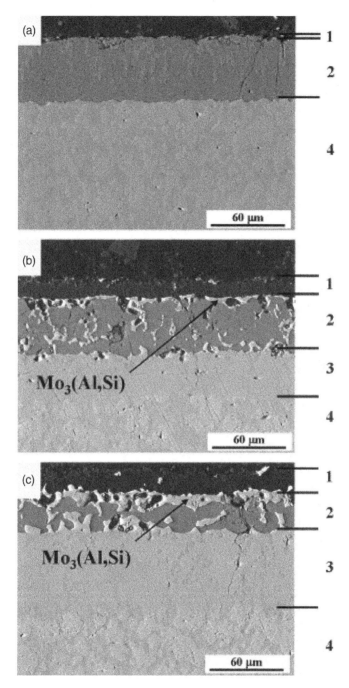

FIGURE 8.6 BSE (Back scattered electron) microstructure of 50 h oxidized pack aluminum coated Mo–3Si–1B alloy at (a) 732°C; (b) 1300°C; (c) 1372°C, Al_2O_3 layer at top (1st layer), Mo_3Al_8 (dark phase) + $Mo_3(Al, Si)$ (2nd layer), $Mo_3(Al, Si)$ (brightest phase) + Mo_5SiB_2 (3rd layer), and substrate phase (4th layer). (Reprinted from [53]. Copyright (2006), with permission from Elsevier.)

the formation of the oxide scale with low solubility of Al impedes the reduction in Al exhaustion and improves the oxidation resistance [54].

Although the fine grain microstructure in ODS refractory alloys can promote oxygen diffusion which may form volatile/non-volatile oxides, the oxidation resistance depends on the continuity of the oxide scale. The protective nature of Cr_2O_3 at the outer layer can be forfeited at temperatures > 1000°C ascribed to volatile CrO_3 formation [55]. Therefore, alloys with an application temperature regime > 1000°C with Cr_2O_3 need compositional modification to improve the high-temperature sustainability. The bonding of Cr_2O_3 with the substrate can be enhanced by the addition of a reactive element so that the $ReCrO_3$ ($YCrO_3$, $LaCrO_3$) formation can reduce oxidation [56]. Nagai and Okabayashi have studied the effect of different levels of oxide dispersion (Y_2O_3, La_2O_3, Al_2O_3, TiO_2, SiO_2, Cr_2O_3) on the elevated temperature oxidation resistance of Ni-20Cr alloys [57]. The investigation reports that mass gain and spalled oxide mass are minimum for La_2O_3 and maximum spallation occurs for SiO_2 dispersion at the oxidation temperatures of 1100°C, 1200°C [57]. The vacancy condensation develops voids and results in inadequate substrate-oxide bonding and hence an oxide dispersion-assisted vacancy sink can be effective to minimize de-bonding of the oxide scale [58]. Further, the improvement in the oxide scale cohesion is attributed to the oxidation reaction responsible for the oxide scale formation, and occurs at the oxide scale-substrate interface in oxide dispersed alloys rather than at the oxide scale-oxygen gas interface in non-oxide dispersed alloys and can enrich the oxide scale cohesion in ODS alloys [59, 60]. The addition of Y_2O_3 has also noticeable significance in restricting the outward cation diffusion at the outer oxide layer and inhibiting the convoluted oxide growth [58]. The morphology of oxide grains is related to the stress generation during scale growth. Development of columnar oxide grains without stress in an alloy with yttrium added can improve the oxide scale bonding compared to stress-assisted equiaxed grain formation [61]. Adhesion of the oxide with the refractory metal is related to the wetting characteristics and it is enhanced by the higher metal-oxygen attraction and the reduction in interfacial energy [62]. The oxide scale cracking and failure may be regarded as due to the lack of thermal shock resistance. Microcracking and toughening by transformation can facilitate superior thermal shock resistance in partially stabilized zirconia (PSZ) which can serve as a potential material for thermal barrier coating (TBC) [63].

8.5 INFERENCES

For high-temperature sustainability of metal/alloys, most importantly the melting point and the oxidation resistance should be sufficient. Several alloying additions such as Zr, Cr, Pd, Y [16, 19, 38, 64] in W, Mo and the application of Al-Si through pack cementation in Nb, Ta [65] contribute to enhancement of oxidation resistance. Apart from alloying additions in elemental form, dispersion of nano-oxide has multiple effects on refractory alloys with respect to oxidation. Reduction in porosity, preventing cationic migration to the outer oxide layer specifically for certain oxides, such as Cr_2O_3, may degrade the passivation at temperatures > 1000°C, which need additional protection. Cracks in the oxide scale can be significant if more oxides form with a major CTE variation than the substrate. Improved flow properties and crack arresting offered by the dispersed oxide add to the advantages of the reduced oxidation rate. The reduced sublimation of oxides is possible by forming a complex mixed oxide through a reaction between the volatile oxides containing the base element and the dispersed oxides. The solubility of oxygen in oxides, the rate of impervious oxide formation during oxidation, the particle pressure of oxygen and the rate of cationic diffusion through the voids, the ionic radius of the oxide-forming element, and the resistance to temperature-induced shock are equally essential to control the oxidation rate of refractory alloys.

REFERENCES

[1] J. R. DiStefano, B. A. Pint, J. H. DeVan, Oxidation of refractory metals in air and low pressure oxygen gas, *Int. J. Refract. Met. Hard Mater.*, 18 (4–5) (2000) 237–243.

[2] O. N. Senkov, J. Gild, T. M. Butler, Microstructure, mechanical properties and oxidation behavior of NbTaTi and NbTaZr refractory alloys, *J. Alloys Compd.*, 862 (2021) 158003.

[3] O. N. Senkov, S. I. Rao, T. M. Butler, T. I. Daboiku, K. J. Chaput, Effect of Fe additions on the microstructure and properties of Nb-Mo-Ti alloys, *Int. J. Refract. Met. Hard Mater.*, 89 (2020) 105221.

[4] T. M. Butler, Oxidation behavior of a refractory alloy C103, AFRL, Wright-Patterson AFB, Dayton, OH, (2020).

[5] K. S. Thomas, S. K. Varma, Oxidation response of three Nb-Cr-Mo-Si-B alloys in air, *Corros. Sci.*, 99 (2015) 145–153.

[6] L. Xiao, X. Zhou, Y. Wang, R. Pu, G. Zhao, Z. Shen, Y. Huang, S. Liu, Z. Cai, X. Zhao, Formation and oxidation behavior of Ce-modified $MoSi_2$-$NbSi_2$ coating on niobium alloy, *Corros. Sci.*, 173 (2020) 108751.

[7] O. N. Senkov, T. I. Daboiku, T. M. Butler, M. S. Titus, N. R. Philips, E. J. Payton, High-temperature mechanical properties and oxidation behavior of Hf-27Ta and Hf-21Ta-21X (X is Nb, Mo or W) alloys, *Int. J. Refract. Met. Hard Mater.*, 96 (2021) 105467.

[8] S. K. Varma, F. Sanchez, S. Moncayo, C. V. Ramana, Static and cyclic oxidation of Nb-Cr-V-W-Ta high entropy alloy in air from 600 to 1400 °C, *J. Mater. Sci. Technol.*, 38 (2020) 189–196.

[9] N. L. Peterson, Diffusion in refractory metals, WADD Technical Report No. 60–793, Wright Air Development Division, Air Research and Development Command, United States Air Force, Wright-Patterson Air Force Base, Dayton OH, (1960).

[10] J. Das, G. A. Rao, S. K. Pabi, M. Sankaranarayana, T. K. Nandy, Oxidation studies on W–Nb alloy, *Int. J. Refract. Met. Hard Mater.*, 47 (2014) 25–37.

[11] B. Li, S. Jiang, K. Zhang, Effects of W content on high temperature oxidation resistance and room temperature mechanical properties of hot-pressing Nb–XW alloys, *Mater. Sci. Eng. A*, 556 (2012) 15–22.

[12] S. Telu, R. Mitra, S. K. Pabi, High temperature oxidation behavior of W–Cr–Nb alloys in the temperature range of 800–1200 °C, *Int. J. Refract. Met. Hard Mater.*, 38 (2013) 47–59.

[13] N. P. Bansal, R. H. Doremus, *Handbook of Glass Properties*, Academic Press, Orlando, FL, (1986).

[14] K. Li, et al., Crack suppression via in-situ oxidation in additively manufactured W-Ta alloy, *Mater. Lett.*, 263 (2020) 127212.

[15] G. Ouyang, P. K. Ray, S. Thimmaiah, M. J. Kramer, M. Akinc, P. Ritt, J. H. Perepezko, Oxidation resistance of a Mo-W-Si-B alloy at 1000–1300 °C: The effect of a multicomponent Mo-Si-B coating, *Appl. Surf. Sci.*, 470 (2019) 289–295.

[16] X. Y. Tan, et al., Evaluation of the high temperature oxidation of W-Cr-Zr self-passivating alloys, *Corr. Sci.*, 147 (2019) 201–211.

[17] W. Ni, L. Liu, Y. Zhang, H. Fan, G. Song, D. Liu, G. Benstetter, G. Lei, Mass loss of pure W, W-Re alloys, and oxide dispersed W under ITER-relevant He ion irradiations, *J. Nucl. Mater.*, 527 (2019) 151800.

[18] M. D. P. Moricca, S. K. Varma, Isothermal oxidation behaviour of Nb–W–Cr Alloys, *Corr. Sci.*, 52 (9) (2010) 2964–2972.

[19] D. B. Lee, G. Simkovich, Oxidation of Mo-W-Cr-Pd alloys, *J. Less Common Met.*, 163(1) (1990) 51–62.

[20] S. Telu, R. Mitra, S. K. Pabi, Effect of Y_2O_3 addition on oxidation behavior of W-Cr alloys, *Metall. Mater. Trans. A*, 46 (2015) 5909–5919.

[21] A. Patra, R. R. Sahoo, S. K. Karak, S. K. Sahoo, Effect of nano Y_2O_3 dispersion on thermal, microstructure, mechanical and high temperature oxidation behavior of mechanically alloyed W-Ni-Mo-Ti, *Int. J. Refract. Met. Hard Mater.*, 70 (2018) 134–154.

[22] D. Sun, C. Liang, J. Shang, J. Yin, Y. Song, W. Li, T. Liang, X. Zhang, Effect of Y_2O_3 contents on oxidation resistance at 1150 °C and mechanical properties at room temperature of ODS Ni-20Cr-5Al alloy, *Appl. Surf. Sci.*, 385 (2016) 587–596.

[23] V.R. Talekar, A. Patra, S. K. Sahoo, Oxidation behavior of oxide dispersion-strengthened W–Ni alloys. *Oxid. Met.*, 93 (2020) 17–28.

[24] E. Lang (Ed.), *The Role of Active Elements in the Oxidation Behaviour of High Temperature Metals and Alloys;* Elsevier Applied Science: London, (1989).

[25] T. B. Reed (Ed.), *Free Energy of Formation of Binary Compounds;* MIT Press, Cambridge, MA, (1971).

[26] R. D. Shannon, Revised effective ionic radii and systematic studies of interatomic distances in halides and chalcogenides, *Acta Crystallogr. Sect. A*, 32 (5) (1976) 751–767.

[27] P. Jehanno, M. Boning, H. Kestler, M. Heilmaier, H. Saage, M. Kruger, Molybdenum alloys for high temperature applications in air, *Powder Metall.*, 51 (2) (2008) 99–102.

[28] J. D. Kuenzly, D. L. Douglass, The oxidation mechanism of Ni_3Al containing yttrium, *Oxid. Met.*, 8 (1974) 139–178.

[29] J. E. Antill, K. A. Peakall, Influence of an alloy addition of yttrium on the oxidation behaviour of an austenitic and a ferritic stainless steel in carbon dioxide, *J. Iron Steel Inst.*, 205 (1967) 1136–1142.

[30] J. M. Francis, J. A. Jutson, High temperature oxidation of an Fe-Cr-Al-Y alloy in CO_2, *Corros. Sci.*, 8 (1968) 445–449.

[31] J. K. Tien, F. S. Pettit, Mechanism of oxide adherence on Fe-25Cr-4Al (Y or Sc) alloys. *Metall. Mater. Trans. B*, 3 (1972) 1587–1599.

[32] T. A. Ramanarayanan, M. Raghavan, R. P. Luton, The characteristics of alumina scales formed on Fe-based yttria-dispersed alloys, *J. Electrochem. Soc.*, 131 (1984) 923.

[33] W. J. Quadakkers, Growth mechanisms of oxide scales on ODS alloys in the temperature range 1000–1100°C, *Materials and Corros.*, 41 (1990) 659–668.

[34] J. H. Koh, E. Sorge, T. C. Wen, D. K. Shetty, Thermal expansion behaviors of yttrium tungstates in the WO_3–Y_2O_3 system, *Ceram. Int.*, 39 (7) (2013) 8421–8427.

[35] H. J. Borchardt, Yttrium-tungsten oxides, *Inorg. Chem.*, 2 (1963) 170–173.

[36] K. S. Chan, Cyclic oxidation response of multiphase niobium-based alloys. *Metall. Mater. Trans. A*, 35 (2004) 589–597.

[37] F. Yang, K. S. Wang, P. Hu, H. C. He, X.Q. Kang, H. Wang, R. Z. Liu, A. A. Volinsky, La doping effect on TZM alloy oxidation behavior, *J. Alloys Compd.*, 593 (2014) 196–201.

[38] S. Telu, A. Patra, M. Sankaranarayana, R. Mitra, S.K. Pabi, Microstructure and cyclic oxidation behavior of W–Cr alloys prepared by sintering of mechanically alloyed nanocrystalline powders, *Int. J. Refract. Met. Hard Mater.*, 36 (2013) 191–203.

[39] W. D. Klopp, Recent developments in chromium and chromium alloys, *NASA-Report TM X-1867* (1969) 1–11.

[40] R. P. Elliot, Columbium-Oxygen system, *Trans. Am. Soc. Met.*, 52 (1960) 990–1014.

[41] H. Jehn, E. Olzi, High temperature solid-solubility limit and phase studies in the system tantalum-oxygen, *J. Less Common Met.*, 27 (3) (1972) 297–309.

[42] N. Birks, G. H. Meier, F. S. Pettit, *Introduction to the High Temperature Oxidation of Metals*, (2nd edn), Cambridge University Press, Cambridge, (2006).

[43] H. J. Maier, T. Niendorf, R. Bürgel, *Handbuch Hochtemperatur-Werkstofftechnik: Grundlagen, Werkstoffbeanspruchungen, Hochtemperaturlegierungen und Beschichtungen*, Springer Verlag, Wiesbaden, (2019).

[44] S. D. Cramer, B. S. Covino Jr., *ASM Handbook*, vol. 13B, *Corrosion: Materials*, ASM International, Materials Park, OH, (2005).

[45] M. G. Frohberg, *Thermodynamik für Werkstoffwissenschaftler, -Ingenieure und Metallurgen: Eine Einführung*, Wiley-VCH-Verlag, Weinheim, (2009).

[46] D. E. Glass, Oxidation and emittance studies of coated Mo-Re, *NASA-CR-201753*, (1997) 1–11.

[47] S. M. C. Fernandes, L.V. Ramanathan, Influence of rare earth oxide coatings on oxidation behaviour of Fe–20Cr alloys, *Surf. Eng.*, 16 (4) (2000) 327–332.

[48] S. M. C. Fernandes, L. V. Ramanathan, Cyclic oxidation behaviour of rare earth oxide coated Fe–20Cr alloys, *Surf. Eng.*, 22 (4) (2006) 248–252.

[49] J. K. Tien, J. M. Davidson, Oxide spallation mechanisms, in: J. V. Cathcart (Ed.), *Stress Effects and the Oxidation of Metals*, AIME, New York, (1975) 200–219.

[50] H. E. Evans, Stress effects in high temperature oxidation of metals, *Int. Mater. Rev.*, 40 (1) (1995) 1–40.

[51] J. Robertson, M. I. Manning, Limits to adherence of oxide scales, *Mater. Sci. Technol.*, 6 (1) (1990) 81–92.

[52] M. V. Aguirre, A. Martín, J. Y. Pastor, J. LLorca, M. A. Monge, R. Pareja, Mechanical properties of tungsten alloys with Y_2O_3 and titanium additions, *J. Nucl. Mater.*, 417 (1–3) (2011) 516–519.

[53] R. Sakidja, F. Rioult, J. Werner, J. H. Perepezko, Aluminum pack cementation of Mo–Si–B alloys, *Scr. Mater.*, 55 (10) (2006) 903–906.

[54] A. S. Ulrich, M. C. Galetz, Protective aluminide coatings for refractory metals, *Oxid. Met.*, 86 (2016) 511–535.

[55] P. Berthod, Kinetics of high temperature oxidation and chromia volatilization for a binary Ni–Cr alloy, *Oxid. Met.*, 64 (2005) 235–252.

[56] S. Chevalier, J. P. Larpin, Formation of perovskite type phases during the high temperature oxidation of stainless steels coated with reactive element oxides, *Acta Mater.*, 50 (12) (2002) 3105–3114.

[57] H. Nagai, M. Okabayashi, High-temperature oxidation of Ni-20Cr alloys with dispersion of various reactive metal oxides, *Trans. Jpn. Inst. Met.*, 22 (2) (1981) 101–108.

[58] F. A. Golightly, F. H. Stott, G. C. Wood, The influence of yttrium additions on the oxide-scale adhesion to an iron-chromium-aluminum alloy, *Oxid. Met.*, 10 (1976) 163–187.

[59] J. Stringer, B. A. Wilcox, R. I. Jaffee, The high-temperature oxidation of nickel-20 wt.% chromium alloys containing dispersed oxide phases, *Oxid. Met.*, 5 (1972) 11–47.

[60] J. Stringer, I. G. Wright, The high-temperature oxidation of cobalt-21 wt.% chromium-3 vol.% Y_2O_3 alloys, *Oxid. Met.*, 5 (1972) 59–84.

[61] F. N. Rhines, J. S. Wolf, The role of oxide microstructure and growth stresses in the high-temperature scaling of nickel, *Metall. Mater. Trans. B*, 1 (1970) 1701–1710.

[62] G. V. Samsonov, R. A. Alfintseva, Dispersion strengthening of refractory metals: A survey, *Powder Metall. Met. Ceram.*, 11 (1972) 98–108.

[63] D. W. Richerson, *Modern Ceramic Engineering: Properties, Processing, and Use in Design*, Marcel Dekker, New York, (1982) 38–45, 139–142.

[64] T. Wegener, F. Klein, A. Litnovsky, M. Rasinski, J. Brinkmann, F. Koch, Ch. Linsmeier, Development of yttrium-containing self-passivating tungsten alloys for future fusion power plants, *Nucl. Mater. Energy*, 9 (2016) 394–398.

[65] S. Majumdar, T. P. Sengupta, G. B. Kale, I. G. Sharma, Development of multilayer oxidation resistant coatings on niobium and tantalum, *Surf. Coat. Technol.*, 200 (2006) 3713–3718.

9 Application-Oriented Strategy

9.1 PROPERTIES THAT INFLUENCE THE APPLICATIONS

The study of ODS-based refractory alloys has advanced significantly due to their interesting microstructure and superior properties required for structural applications. In previous chapters, the individual properties of ODS refractory alloys with respect to final applications have been discussed. Briefly, all the properties for diverse applications of some ODS refractory alloys are presented in Table 9.1.

The major ODS refractory grade alloys as presented in Table 9.1 also indicate that there is immense potential for diverse applications of the alloys compared to non-ODS alloys. The majority of the research work on ODS alloys is centered on ODS steel. However, interest has developed in several excellent properties of ODS refractory alloys required for critical applications, which can be an alternative to traditional alloys. For example, Nb-based alloys attract a lot of attention rather than Ni-based superalloys for turbine blade applications. W-based alloys have replaced depleted uranium (DU) with respect to kinetic energy penetrator applications, owing to the problems related to the environment and health [14, 15]. It is also reported that adiabatic shear band formation leads to self-sharpening in kinetic energy penetrator applications, though a divergent mode of fracture (interfacial fracture) in 93W-4.9Ni-2.1Fe-0.1wt.% Y_2O_3 alloy facilitates self-sharpening behavior and improves the efficiency of the kinetic energy penetrator [1]. During the kinetic energy penetrator application, due to high impact and elevated kinetic energy, a significant temperature rise is obvious. Therefore, the high melting point of the dispersed oxide when added in W-based alloys, the density of the alloys, and the oxidation resistance for significant penetration are the major aspects for attention. Kang and Park have studied the effect of using copper infiltrated tungsten (CIT) and W/Y_2O_3 as nozzle throat inserts and reported that the rate of ablation of W/Y_2O_3 is 55% compared to CIT [4]. Lower oxidation and wear improve the sustainability for both kinetic energy penetrators and rocket nozzle throats. $Nb-TiO_2$ alloy has also been reported as a prospective entrant in medical implants applications and the selection of Nb over Ta is connected to its superior density and cost effectiveness [12]. High wear resistance coupled with strength offered by dispersed oxide contributes to enhancing durability for biomedical applications. As discussed for the properties of high temperature applications of Nb-based ODS alloys, the dispersoids' coarsening needs to be controlled. Advances in the fabrication of superalloys from conventional Ni-based to Nb-based ODS dispersion have strengthened alloys aimed at improving the high temperature strength. The agglomeration tendency of nano-dispersoids added in refractory metals/alloys during the synthesis phase can be regulated if the sintering temperature and the time are minimized or the heating rate is enhanced, as in the spark plasma sintering

TABLE 9.1
Applications of ODS Refractory Alloys and Associated Properties

ODS refractory alloys	Applications	Application-based properties
93W-4.9Ni-2.1Fe-0.1wt.% Y_2O_3 [1] 94W-3.65Ni-0.91Fe-1.14Mo-0.3Y_2O_3 and 94W-4.56Ni-1.14Co-0.3Y_2O_3 [2]	Kinetic energy penetrator	(i) Reduced contiguity of W/matrix (ii) Shear localization/Adiabatic shear (iii) Homogenization of dispersed oxides (iv) Oxide dispersion inside the matrix (v) High melting point (vi) High density (vii) High erosion, oxidation resistance at elevated temperature.
W-1% La_2O_3 [3]	Plasma-facing material in nuclear reactor	(i) Refined grain size (ii) High thermal shock resistance (iii) Impeding recrystallization at the application temperature (iv) Uniform dispersion of oxides (v) Higher erosion resistance
W-Y_2O_3 [4]	Rocket nozzle throat	(i) High melting point (ii) Dimensional precision: low thermal expansion coefficient [5] (iii) Low erosion/ablation rate (iv) Thermal shock resistance [5]
W-1 wt.%La_2O_3 [6]	Electrodes for plasma spraying	(i) Low electrical power (ii) Reduced cathode temperature (iii) Reduced wear (iv) Reduced work function (v) Decreased reduction rate of arc voltage
Mo-La_2O_3 [7–10]	Furnace heating elements, Sintering boats, lamp components	(i) Resistance against creep deformation (ii) Elevated recrystallization temperature (iii) Improved percentage elongation (iv) Enhanced high temperature stability
Mo-ZrO_2 [9]	Glass-melting furnace components	(i) Enhanced current density (ii) Superior strength (iii) Hindrance to thermal shock (iv) High temperature withstand capacity (1700°C), (v) Enriched durability
Mo-Y_2O_3-Ce_2O_3 [9,11]	Halogen lamp	(i) Improved high temperature strength (ii) Enhanced mechanical stability (iii) Corrosion resistance to halogen gas in the lamp

TABLE 9.1 (Continued)
Applications of ODS Refractory Alloys and Associated Properties

ODS refractory alloys	Applications	Application-based properties
Nb-1%TiO$_2$ [12]	Medical implants	(i) Load-bearing capacity (ii) High strength (iii) Corrosion resistance to body fluids
Nb-Al-Cr-V-Ti-(Al$_2$O$_3$/Y$_2$O$_3$) [13]	Heat-resistant alloys	(i) Resistance against high temperature oxidation (ii) Adequate strength at high temperature (iii) Low solubility of oxygen in Nb at elevated temperature (iv) Resistance to dispersion growth

technique. Research on Ta-Yttrium nitride (YN) and Ta-Y$_2$O$_3$-Si with respect to the application of wire products shows that the latter produces more grain refinement, which is a prerequisite in preventing embrittlement [16]. Furthermore, the enhanced stability of oxide dispersoids also leads to the following: (1) it causes a higher strain development in the lattice compared to nitrides; and (2) it impedes the dissolution of the dispersoids in the matrix [16]. The selection of W, as plasma-facing material (PFM), is due to the elevated entry value of hydrogen sputtering energy, however, W exhibits a significant power drop while exposed to plasma, owing to its high atomic number [17]. Mo possesses a lower atomic number than W but is used in regions of limiter and wall without any impurity formation in the high-density areas [17]. The literature shows a higher reduction in thermal conductivity with temperature for W-1% La$_2$O$_3$ compared to W, and also reports significant cracking of W-1% La$_2$O$_3$ in contrast to W, when subjected to edge plasma loading, which leads to high surface roughness in W-1% La$_2$O$_3$ [18]. It is concluded in the paper that the wt.% La$_2$O$_3$ addition needs to be optimized to control the drop in the melting point of the alloys, due to the lower melting temperature of La$_2$O$_3$ against W, as well as the nano-size of La$_2$O$_3$ should be used in the areas of small momentary events in the fusion reactor [18]. Refined oxide particles with a higher fraction incorporate a strong resistance to grain growth, increase thermal activation for the movement of the grain boundary and enhance the recrystallization temperature [19]. The distance of the nano-oxide in the material influences the extent of the damage of W-ODS alloys for PCM application under the radiation effect. The density of Y$_2$O$_3$ is considerably smaller (5.01 g/cm^3) than W (19.3 g/cm^3) and therefore the motion path of α particles impacted on the nano-oxide does not diverge from its initial path [20]. The literature data shows that the Y$_2$O$_3$ nano-particles can inhibit the α particles more effectively than the W matrix [21, 22], however, the atomic shift is significantly less in Y$_2$O$_3$ nano-particles than the W matrix [20]. It is also reported that the location of nano-oxide at a small distance from the surface develops more impairment at a higher depth from the surface [20].

9.2 IMPROVEMENT OF APPLICABILITY

The application regime of refractory-based ODS alloys depends on several properties, as described in Section 9.1. To achieve the desired properties, microstructural tailoring and control of the process parameters and selection of a suitable processing route are of the utmost importance. The challenge to retain the nano-scale dimension of dispersed oxide to improve the mechanical properties can be addressed by limiting the sintering time, the temperature and increasing the heating rate. During the processing of ODS refractory alloys by an advanced sintering process, such as SPS, significant temperature deviation between outer die wall and inner material may lead to improper densification which deteriorates the mechanical properties. Moreover, a sudden temperature rise or cooling may cause a thermal shock, followed by cracking and therefore resistance to thermal shock is a prerequisite for several applications. The resistance can be achieved by suitably designing the sintering cycle and controlling the grain size as too high a coarse grain and an ultrafine grain disrupt the thermal shock resistance [23]. The creep resistance of ODS refractory alloys at high temperature applications also are of immense importance. This can be accomplished through restricting the grain boundary and dislocation movement by dispersed oxides [24]. An increase in application temperature causes the drop in hardness that may influence the wear resistance. Additionally, during high temperature exposure, the erosion/wear property of alloys depends on the adhesion of the oxide scale to the matrix, the porosity of the alloy, and the fracture toughness of the oxide scale. It is reported that oxidative wear and chipped off oxide particles provide lubrication and improve the wear resistance of the alloy, however, the wear resistance reduces with the increase in hardness of the oxides and the roughness of the alloy [25]. The enhancement of the applicability of pure W with respect to fusion reactor applications by oxide dispersion is accredited to the improved swelling resistance and elevated temperature strength, and in high temperature applications, the creep rupture strength needs to be improved, which is made possible by the dispersion of fine oxides [26]. Additionally, coarse grain obtained by prolonged high temperature exposure or in the sintering stage enhances the hindrance to creep but large grains with high aspect ratio increase the anisotropic properties [26]. Therefore, the application temperature regime should be less than the recrystallization temperature of the alloy, and oxide doping in the intragranular region is an excellent strategy to optimize the strength and ductility. The load-bearing capacity of the matrix improves by adding hard oxide and the oxide can impede the load applied on the material, reducing the wear loss [27], though the capacity depends on the magnitude of the load [28]. High hardness of W is a factor for wear resistance applications, however, due to the high DBTT of W, the wear resistance suffers. It is reported that the addition of stable nano-oxide can contribute to improving the wear resistance of the alloy [27, 28]. The rate of volatilization of refractory oxides, such as WO_3, MoO_3, Re_2O_7, is a deciding factor for oxidation resistance at a high temperature [29]. The literature reports that the addition of Re to W improves the oxidation resistance but still a low volatilization temperature of Re_2O_7 [29] and the cost of Re are a concern. Reports suggest the improvement of oxidation resistance of W-Re alloy wire by the presence of La oxide to enhance the application regime [30].

Based on the discussion in Sections 9.1 and 9.2, it is obvious that the addition of oxide dispersion to regulated processing can improve the limitations of refractory metals alloys, which lead to failure of components in service, and therefore this enriches applicability. The rare earth oxides are costly and adding a minimal quantity is advantageous with respect to overall product cost and the desired mechanical properties.

REFERENCES

[1] S. Park, D. K. Kim, S. Lee, D. K. Kim, H. J. Ryu, S. H. Hong, H. J. Ryu, Dynamic deformation behavior of an oxide-dispersed tungsten heavy alloy fabricated by mechanical alloying, *Metall Mater Trans A*, 32 (2001) 2011–2020.

[2] S. H. Hong, H. J. Ryu, S. I. Cha, H. Y. Kim, K. T. Kim, K. H. Lee, C. B. Mo, Development of advanced oxide dispersion strengthened tungsten heavy alloy for penetrator application, Accession Number: ADA474421, Korea Advanced Institute of Science and Technology, Daejeon, Korea, (2005) 1–47.

[3] J. W. Davis, V. R. Barabash, A. Makhankov, L. Plochl, K. T. Slattery, Assessment of tungsten for use in the ITER plasma facing components, *J. Nucl. Mater.*, 258–263 (Part 1) (1998) 308–312.

[4] Y. G. Kang, J. H. Park, Thermal performance evaluations of tungsten/yttria as nozzle throat insert material for long duration firing, *J. Korean Soc. Aeronaut. Space Sci.*, 38 (2) (2010) 200–205.

[5] A. Upadhyaya, P/M alloys for aerospace applications, in: P. Ramakrishnan (Ed.), *Powder Metallurgy*, New Age International, New Delhi, (2007) 38.

[6] M. Heißl, C. Mitterer, T. Granzer, J. Schröder, M. Kathrein, Substitution of thoria additions by lanthanum-oxide doping in electrodes for atmospheric plasma spraying, 18th Plansee Seminar, (2013).

[7] Plansee, Refractory metal heating elements. Available at: www.plansee.com/en/products/components/furnace-construction/heating-elements.html (accessed August 2, 2021).

[8] Plansee, Metallic hot zones from the experts. For your new furnace or as a replacement. Available at: www.plansee.com/en/products/components/furnace-construction/hot-zones.html (accessed August 2, 2021).

[9] J. A. Shields, *Applications of Molybdenum Metal and its Alloys*, (2nd edn), International Molybdenum Association (IMOA), London, (2013).

[10] P. M. Cheng, G. J. Zhang, J. Y. Zhang, G. Liu, J. Sun, Coupling effect of intergranular and intragranular particles on ductile fracture of Mo–La_2O_3 alloys, *Mater. Sci. Eng. A*, 640 (2015) 320–329.

[11] G. Leichtfried, G. Thurner, R. Weirather, Molybdenum alloys for glass-to-metal seals, *Int. J. Refract. Met. Hard Mater.*, 16 (1) (1998) 13–22.

[12] U. Gennari, E. Kny, T. Gartner, Niobium-titanium oxide alloy: A novel approach to oxide dispersion strengthening, in: H. Bildstein, H. M. Ortner (Eds.), *Proceedings of the 12th International Plansee Seminar, Reutte, May 1989*, vol. 3, Verlagsanstalt Tyrolia, (1989), 587–614.

[13] K. Tsukuta, C. T. Iikubo, Oxide-dispersion-strengthened niobium-based alloys and process for preparing, United States Patent, Patent Number: 5,180,446, (1993).

[14] A. Upadhyaya, Processing strategy for consolidating tungsten heavy alloys for ordnance applications, *Mater. Chem. Phys.*, 67 (1–3) (2001) 101–110.

[15] M. C. Hogwood, A. R. Bentley, The development of high strength and toughness fibrous microstructures in tungsten-nickel-iron alloys for kinetic energy penetrator applications,

in: A. Bose, R. J. Dowding (Eds.), *Tungsten and Refractory Metals*, Metal Powder Industries Federation, Princeton, NJ, (1994) 37–45.

[16] P. Kumar, C. E. Mosheim, Effect of intermetallic compounds on the properties of tantalum, *Int. J. Refract. Met. Hard Mater.*, 12 (1) (1993) 35–40.

[17] M. Ulrickson, V. Barabash, S. Chiocchio, G. Federici, G. Janeschitz, R. Matera, M. Akiba, G. Vieider, C. Wu, I. Mazul, Selection of plasma facing materials for ITER, in: *Proceedings of 16th International Symposium on Fusion Engineering*, vol. 1 (1995) 394–398.

[18] D. H. Zhu, J. L. Chen, Z. J. Zhou, R. Yan, R. Ding, Influences of dispersed lanthanum oxide additive on the properties of tungsten-based plasma-facing material, *Fusion Sci. Technol.*, 66 (2) (2014) 337–342.

[19] T. Zhang, Z. Xie, J. Yang, T. Hao, C. Liu, The thermal stability of dispersion-strengthened tungsten as plasma-facing materials: A short review, *Tungsten*, 1 (2019) 187–197.

[20] M. G. Petaccia, J. L. Gervasoni, Nano-yttria in oxide dispersion strengthened tungsten under alpha particle irradiation, *Nucl. Mater. Energy*, 20 (2019) 100681.

[21] Anon, Report 90, *J. Int. Commiss. Radiat. Units Meas.*, 14 (1) (2016), https://doi.org/10.1093/jicru/ndw043.

[22] C. Williamson, J. Boujot, J. Picard, Tables of range and stopping power of chemical elements for charged particles of energy 0.5 to 500 MeV, Technical Report, CEA-R--3042, Commissariat à l'Enérgie Atomique, (1966).

[23] G. Korb, M. Ruhle, H. P. Martinez, New iron-based ODS-superalloys for high demanding applications, in: *Proceedings of the ASME 1991, International Gas Turbine and Aeroengine Congress and Exposition*, vol. 5: *Manufacturing Materials and Metallurgy; Ceramics; Structures and Dynamics; Controls, Diagnostics and Instrumentation; Education; IGTI Scholar Award; General*, ASME, Orlando, FL, (1991) V005T12A015.

[24] P. Susila, D. Sturm, M. Heilmaier, B. S. Murty, V. S. Sarma, Effect of yttria particle size on the microstructure and compression creep properties of nanostructured oxide dispersion strengthened ferritic (Fe-12Cr-2W-0.5Y_aO_a) alloy, *Mater. Sci. Eng. A*, 528 (13–14) (2011) 4579–4584.

[25] W. B. Rowe, Mechanics of abrasion and wear, in: W. B. Rowe (Ed.), *Principles of Modern Grinding Technology*, (2nd edn), William Andrew Publishing, Norwich, (2014) 349–379.

[26] M. S. El-Genk, J. M. Tournier, A review of refractory metal alloys and mechanically alloyed-oxide dispersion strengthened steels for space nuclear power systems, *J. Nucl. Mater.*, 340 (2005) 93–112.

[27] G. Cui, Y. Liu, S. Li, H. Liu, G. Gao, Z. Kou, Nano-TiO_2 reinforced CoCr matrix wear resistant composites and high-temperature tribological behaviors under unlubricated condition. *Sci. Rep.*, 10 (2020) 6816.

[28] P. D. Srivyas, M. S. Charoo, Tribological behavior of aluminum silicon eutectic alloy based composites under dry and wet sliding for variable load and sliding distance, *SN Appl. Sci.*, 2 (2020) 1654.

[29] T. Otsuka, N. Sawano, Y. Fujii, T. Omura, C. Taylor, M. Shimada, Effects of rhenium contents on oxidation behaviors of tungsten-rhenium alloys in the oxygen gas atmosphere at 873 K, *Nucl. Mater. Energy*, 25 (2020) 100791.

[30] X. Bo, Q. Liu, H. Yang, C. Jiang, H. He, X. Wang, C. Liu, D. Chen, The research on improving the anti-oxidation of tungsten rhenium alloy wires, *J. Phys.: Conf. Ser.*, 885 (2017) 012019.

10 Future Potential

10.1 TREND TOWARD IMPROVEMENT OF SUSTAINABILITY

In any application, the major objective is to use the material for a longer life span or improve its durability. It is also true that repeated applications of materials at high temperatures involving a stress field cause thermal or mechanical stress to develop and if the residual stress relaxation has not been achieved in each cycle, the stress magnitude keeps on increasing, which may also trigger sudden failure. It is manifested that the addition of ceramic oxide particles adds to the stress concentration in the material at the grain boundary region and can exaggerate the crack propagation. Enhancement of sustainability can be accomplished by minimizing the limitations in both low and high temperature regimes. In a rapid spectrum reactor, ferritic steels with low activation are produced, which are identical to ferritic-martensitic steels [1]. The need for enhanced durability and a radiation effect for a longer time period led to the development of ODS alloys [1]. The major concerns for refractory metals and alloys are: (1) creep properties; (2) toughness; (3) embrittlement caused by radiation; (4) corrosion; (5) oxidation; and (6) the presence of impurities [2]. Several refractory-based high entropy alloys have been developed with the aim of improving the high temperature strength and effect on the environment [3]. The cracking in applications can be counteracted by self-healing or crack-closing mechanisms. Alloys can also self-passivate during high temperature exposure to reduce further weight loss. For residual stress relaxation, there should be sufficient plastic deformation and this depends on the available slip systems. The ambient temperature ductility of BCC-refractory high entropy alloys (HEA) is limited and can be improved by suitably modulating the grain size and processing conditions [4]. In high entropy alloys, the cost and availability of individual elements need to be taken into account. Several refractory elements are also recovered during the iron or steel, Al, Cu, Zn or Pb extraction process and can add to the availability of the elements [5]. The market competitiveness depends on the Herfindahl-Hirschman Index (HHI), which indicates that W, Nb, Ru, Rh, Y have increased, therefore, the competition is less in contrast to Mo, Hf, Ta-based industries [5]. It is reported that doping of interstitial elements in BCC-refractory alloys reduces ductility, but the modification of the constituents to form HEA of TiZrHfNb alloy with oxygen doping promotes strength and ductility, owing to the variation in the deformation method (planar slip → wavy slip) by the oxygen short-range order [6, 7]. The requirement of sustainable material, such as refractory grade coatings, has significant potential in the corrosive condition of a reactor [1]. The production of nuclear waste undergoes stringent environmental norms and the materials employed must be of minimum activation [1]. The sustainability of materials also depends on: (1) cost effectiveness; (2) safe application; and (3) reduced generation of nuclear waste, therefore, research on oxide dispersion strengthened refractory alloys is a step forward in that direction. Several new alloys such as Nb-ODS are being developed for medical implants apart from commonly used Ti-based alloys. The requirement of improved biocompatibility, closer elastic modulus to bone, and wear resistance of Nb alloys shows a pathway for the same kind of applications. The creep rupture strength is a prime factor to evaluate the high temperature durability of alloys. The creep rupture strength improves considerably by oxide dispersion in Mo, even though the creep rupture strength of ODS-Mo is higher at 1800°C than W, Re at 1600°C [8]. Additionally, the DBTT processing condition also is related to a sustainable application regime of ODS alloys. The DBTT of ODS-Mo with Re is highly negative in the swaged condition compared to the recrystallized condition [9], and the same is true also without the Re addition [8]. The aim of using refractory alloys for high temperature application is to achieve creep stress of 70–138 MPa with a corresponding temperature and time of 1200°C and 10,000 h respectively [10]. Commercial alloys (TZM) exhibit appreciable creep prevention at 1100°C, however, with an increase in temperature to 1200°C, the creep stress is much less (21 MPa) [11]. Oxide dispersion can improve the creep stress to 103 MPa at 1200°C in contrast to TZM alloy [9]. For a gas-cooled fast reactor (GFR) which uses He gas as a coolant and has core outlet temperature of around 850°C, it is reported that refractory alloys are the first choice [2]. In such reactors, the retained oxygen is more influential rather than the corrosion effect [2]. A comparison has been made among various refractory metals and alloys (Nb-1Zr, Ta-10W, Mo-0.5Ti-0.1Zr, W-Re, Re) with respect to properties [12]. The results indicate that Nb-1Zr is best in fabricability, Nb-1Zr, Ta-10W, Re are identical in weldability, Re exhibits superior creep strength and oxidation resistance is exhibited in Ta-10W, and Mo-0.5Ti-0.1Zr, W-Re show superior compatibility with alkali metal [12]. Though the results show encouraging prospects of Re in terms of high temperature properties (creep strength and oxidation resistance), the associated high cost factor is not justified for its extended applications. The development of strengthened oxide dispersion can counterbalance the limitation of some alloys and enable them to be used in high temperature structure applications. The greater the difference between densities, the melting point of refractory metals and oxides, the most effectively they can be fabricated by the powder metallurgical route and the major aims of refractory alloy fabrication are: (1) improved purity; (2) low cost; (3) decrease in environmental pollution; (4) larger product size; (5) finer microstructure; and (6) enhanced thermal stability [13]. As discussed earlier, purity during fabrication dictates the final properties, and the electrolysis synthesis method can result in improved purity, though the involvement of higher cost and the constraint to produce only elemental powders are the major challenges in this process [13]. Therefore, mechanical alloying of produced elemental powders can develop alloy powders with a refined particle size. Purity can also be warranted by using an inert atmosphere or a vacuum during the alloy powder production and sintering. Morales et al. have reported that reduction by hydrogen of refractory alloys facilitates bulk product fabrication without environmental pollution [13]. Cost effectiveness and improved material yield, with near net shape, come by adopting powder metallurgy rather than vacuum melting, forging, and the machining route for the fabrication of the turbine disk [14, 15]. However, the cost of oxide is a challenge to minimize the overall component cost, but mechanical alloying and bulk-scale fabrication are potential benefits in this context [16].

10.2 FOCUS TO EXTEND THE APPLICATION AREAS

Several application areas of ODS refractory alloys are presented in Chapter 9 (Table 9.1). The ODS refractory alloys have succeeded in competing with non-ODS refractory metals/alloys of other alloys in diverse applications. Superalloys (Ni-based) are mainly used in gas turbine engines. However, there is a constant effort to improve the application temperature in such applications. Improvement in the application temperature range also enhances the efficiency of the gas turbine combined with thermal barrier coatings (TBC) [17]. The research on high temperature alloys is very successful in decreasing fuel consumption [18]. The density of Nb is marginally less than Ni, and the low weight component added to the high temperature strength, fabricability, and efficient weldability lead to the development of Nb alloys (C-103) (Nb-Hf-Ti) [19]. The production of hydrogen from ammonia by decomposition uses Ru/La_2O_3 with high effectiveness [20]. TBC to protect engine components is another important research area and different properties are anticipated: (1) elevated melting point; (2) non-reactive areas; (3) less thermal conductivity; (4) elevated thermal expansion; (5) corrosion resistance at high temperatures; (6) resistance to erosion; and (7) coherence with substrate [21]. Carbides, nitrides, and borides are also candidates along with oxides. The melting point of oxides such as Al_2O_3, ZrO_2, HfO_2, SiO_2, 8YSZ is less than carbides (HfC, TaC, NbC, ZrC, SiC, TiC), nitrides (HfN, ZrN, TaN) and borides (HfB_2, ZrB_2, TiB_2, TaB_2), but the thermal expansion coefficient of Al_2O_3, 8YSZ is either comparable or higher, and the thermal conductivity of ZrO_2, HfO_2, SiO_2, 8YSZ is less than carbides, nitrides and borides [22]. Considering the factor of reducing CO_2 and NO_x and cost effectiveness, rare earth oxide combined with ZrO_2 is a TBC alternative [23]. Extension of the application areas of ODS refractory alloys requires the development of novel processing routes with enough strength and ductility at high temperatures. The literature reports that oxide can be dispersed in a matrix by in-situ or ex-situ methods and the method of dispersion also leads to variation in coherency of dispersoids with the matrix [24, 25]. The main objectives of achieving sustainability in the application of ODS refractory alloys depend on the coherence between oxides and the matrix, the size of the oxides particles after fabrication and the intragranular dispersion of the oxide [26] to counteract any premature failure of the component under load. Moreover, dispersion uniformity during processing by powder metallurgy can be achieved, compared to the melting-casting technique, owing to the difference between oxides and the refractory-based matrix [24]. According to several studies on tensile strength and percentage elongation to failure of W foil/plates [27, 28], W heavy alloys [29, 30], solution-strengthened W alloys [31], and second-phase strengthened W alloys [32, 33], it is shown that tensile strength is higher in W foil/plates and percentage elongation to failure is higher in W heavy alloys, however, the context of low thickness of W foil/plates does not warrant high temperature industrial usage [26]. Using low density alloying elements in ODS refractory alloys is also a prominent strategy to enhance the strength-to-weight ratio by selecting the suitable synthesis and consolidation parameters. Additive manufacturing (AM) or 3D printing is also rapidly booming with respect to producing complex structures through layer-based deposition of powders. The fabrication of refractory-based ODS alloys by the AM technique is not extensive and, therefore,

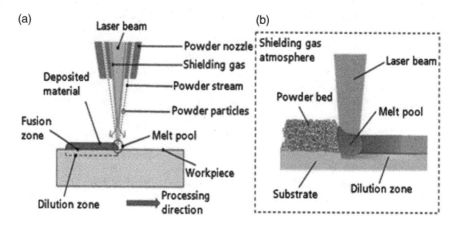

FIGURE 10.1 Schematic of (a) LMD and (b) SLM process. (Reprinted from [42]. Copyright (2018), with permission from Elsevier.)

significant research is needed to reveal the potential of AM. Refractory high entropy alloys (MoNbTaW) can be fabricated by the direct metal deposition technique [34]. Several candidate additive manufacturing methods have been developed, such as powder bed fusion (PBF), binder jetting, metal material extrusion and directed energy deposition, and each process has its own advantages and challenges with respect to accuracy, surface roughness, or internal porosity [35, 36]. The internal porosity is quite low (0.09–0.2%) in directed energy deposition, however, the surface roughness is quite high (20–500 μm) than other processes [35, 37, 38]. Laser metal deposition (LMD) and selected laser melting (SLM) techniques are used [39–41] to fabricate ODS alloys through melt pool formation, as displayed in Figure 10.1 [42]. Reports indicate that LMD and SLM methods can generate uniform oxide dispersion with enriched high temperature properties [42], though oxide contamination during the processing is a concern that can be counteracted by a hot rolling operation to reduce porosity and distort the oxide followed by distribution [43].

Y_2O_3 dispersed Mo-Si-B alloy developed by laser-based additive manufacturing (directed energy deposition) exhibits no cracking and uniform dispersal of Y_2O_3 particles, however, porosity of around 2.1% also raises concerns about high temperature applications [44]. The recent research on $Zr+Er_2O_3$, $Zr+Y_2O_3$ added W alloys provides an interesting comparative insight on the effect of oxide dispersion on the strength of the alloys against other strengthening (grain boundary strength or Peierls–Nabarro stress) [45]. A minimum grain size of 0.58±0.03 μm has been reported for W-Zr-Er_2O_3, owing to the refined size, the elevated population of $Er_2Zr_2O_7$ oxides compared to $Y_2Zr_2O_7$ and the contribution of the grain boundary strengthening is much higher in W-Zr-Er_2O_3, directed at improving the grain size reduction [45].

Though several innovations have occurred in the field of additive manufacturing, certain limitations, such as: (1) the presence of porosities; (2) the incidence of residual stress; (3) anisotropic properties; (4) additional steps after fabrication; (5) inadequate load-carrying ability; (6) defects during solidification [46]; and (7) development of

Future Potential

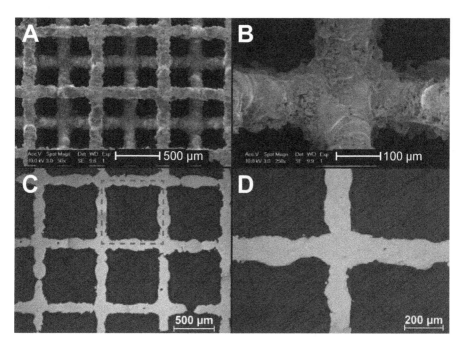

FIGURE 10.2 Structure of porous Ta developed by SLM method, (A, B) SEM image (from top) with variation in magnification; (C, D) Light optical microscopy of cross-section with varied magnification, The sizes are shown in A, C as dotted lines. (Reprinted from [51]. Copyright (2015), with permission from Elsevier.)

micro-cracks have directed attention toward other processes, such as: (1) friction surfacing; (2) friction stir additive manufacturing; (3) friction deposition; (4) friction assisted seam welding; and (5) additive friction stir deposition [47]. The additive friction stir deposition (AFSD) can be applied to achieve an optimized strength-ductility and to regulate the DBTT [48]. W-based components have been fabricated by selective laser melting (SLM), which produces appreciable density (97.8%), however, properties such as microhardness and wear resistance depend on laser fluency, as a higher laser fluency reduces these properties [49]. Laser Engineered Net Shaping (LENS™) is used to develop Ta for bone implants application with 27–55% porosity, Young's modulus of 2–20 GPa, and 0.2% proof strength of 100–746 MPa, and these values are identical to cortical bones in humans [50]. SLM has also been used to fabricate high porosity (80%) (as represented in Figure 10.2) Ta for implants application and the evaluated yield strength and elastic modulus are 12.7 MPa, 1.22 GPa and the properties are in between those of cancellous bones in humans [51].

Several applications of porous Ta in the medical industry, such as in primary femoral stems, complete shoulder components and salvage prostheses, wrist arthrodesis as well as new applications, such as bone graft alternatives, are indicated in the literature [52]. Although the added oxides can act as load-bearing elements in implants applications, research on porous ODS refractory alloys in the domain of medical implants is quite limited.

FIGURE 10.3 Picture of multi-metallic SRF cavity, cavity length (from flange to flange: 31.8 cm, electroplated Cu ring (at equator) diameter is 25.4 cm, thickness is 1.27 cm [58].

C103 products used in propulsion and manufactured using a conventional technique are challenging owing to the high cost, and the complexity in machining [53]. Laser-PBF AM facilitates complex shape fabrication, has enhanced properties, improved yield, reduced buy-to-fly ratio (1.1: 1) against 20: 1-50: 1 in conventional methods for C103 [53].

Fabrication of next generation superconducting radiofrequency (SRF) cavities for linear accelerators from Nb of high purity are fabricated by the electron beam melting additive manufacturing method and the achieved density is 99.7% [53]. The process is capable of producing innovative designs and contamination-free products, which can subdue the Lorentz forces responsible for deformation-assisted detuning [53–56]. Nb plates are also used as tuning plates for superconducting cavities [57]. SRF cavities are also produced from Nb by electroplating Cu at the outer layer to enhance thermal stability and Nb_3Sn at the inner layer as depicted in Figure 10.3 [58].

The process route to develop ODS alloys has also undergone several changes from melting casting to powder metallurgy and, at present, additionally 3D printing/AM, with respect to stringent property requirements and environmental pollution control. The applicability of refractory alloys as functionally graded materials (FGM) of Ta deposition on Mo and Mo on W has been carried out by Wire and Arc Additive Manufacturing [59]. In such a structure, the variation in chemical composition needs to be steady in different desired directions along with gradient interface to improve toughness, reduce residual stress and increase sustainability [60, 61]. More research on FGM (novel refractory alloy design) needs to be carried out to understand its applicability in critical conditions [59]. ODS refractory alloys research studies are mainly concentrated on the development of plasma-facing materials and defense applications, whereas several other areas such as medical implants, rocket nozzles, or furnace components need extensive research with varied dispersions. Exploration of creep rupture life, oxidation behavior, wear properties in such applications will provide a better understanding to tailor the alloy and the process route design.

REFERENCES

[1] B. Raj, M. Vijayalakshmi, P. R. V. Rao, K. B. S. Rao, Challenges in materials research for sustainable nuclear energy, *MRS Bull.*, 33 (2008) 327–337.

[2] Generation IV Roadmap, Crosscutting fuels and materials: R&D scope report, Nuclear Energy Research Advisory Committee and the Generation IV International Forum, (2002).

[3] O. N. Senko, D. B. Miracle, K. J. Chaput, J. Couzinie, Development and exploration of refractory high entropy alloys: A review, *J. Mater. Res.*, 3 (19) (2018) 3092–3128.

[4] O. N Senkov, G. B Wilks, D. B. Miracle, C. P. Chuang, P. K Liaw, Refractory high-entropy alloys, *Intermetallics*, 18 (9) (2010) 1758–1765.

[5] X. Fu, C. A. Schuh, E. A. Olivetti, Materials selection considerations for high entropy alloys, *Scr. Mater.*, 138 (2017) 145–150.

[6] Z. Lei, et al., Enhanced strength and ductility in a high-entropy alloy via ordered oxygen complexes, *Nature*, 563 (2018) 546–550.

[7] J. L. Cann, et al., Sustainability through alloy design: Challenges and opportunities, *Prog. Mater. Sci.*, 117 (2021) 100722.

[8] R. Bianco, R, W. Buckman Jr., Mechanical properties of oxide dispersion strengthened (ODS) molybdenum. In A. Crowson et al. (Eds.), *Molybdenum and Molybdenum Alloys*. TMS, Materials Park, OH, (1998) 125–144.

[9] A. J. Mueller, R. Bianco, R. W. Buckman, Jr., Evaluation of oxide dispersion strengthened (ODS) molybdenum and molybdenum-rhenium alloys, Report No. B-T-3148, U.S. Department of Energy, West Mifflin, Pennsylvania, PA, (1999).

[10] F. Ritzert, M. V. Nathal, J. Salem, N. Jacobson, J. Nesbitt, Advanced stirling duplex materials assessment for potential venus mission heater head application, paper presented at 9th Annual International Energy Conversion Engineering Conference (IECEC), (2011).

[11] R. H. Cooper, E. E. Hoffman, Refractory alloy technology for space nuclear power applications, paper presented at Symposium on Refractory Alloy Technology for Space Nuclear Power Applications, Oak Ridge, TN, USA, (1984).

[12] S. Zinkle, J. T. Busby, K. J. Leonard, L. L. Snead, D. T. Hoelzer, T. S. Byun, paper presented at Embedded Topical Meeting – Nuclear Fuels and Structural Materials for the Next Generation Nuclear Reactors, ANS 2006 Annual Meeting, Reno, NV, USA, (2006).

[13] R. Morales, R. E. Aune, O. Grinder, S. Seetharaman, The powder metallurgy processing of refractory metals and alloys, *JOM*, 55 (2003) 20–23.

[14] J. W. Semmel, Jr., Opportunities in materials and processes for aircraft/ship propulsion gas turbines, paper presented at Materials on the Move, SAMPE National Technical Conference, vol. 6, October (1974).

[15] C. P. Blankenship, Trends in high temperature materials technology for advanced aircraft turbine engines, *SAE Trans.*, 84 (1975) 2905–2916.

[16] J. S. Benjamin, Dispersion strengthened superalloys by mechanical alloying, *Metall Mater. Trans. B*, 1 (1970) 2943–2951.

[17] D. Bonaquist, A. Feuerstein, L. Buchakjian, P. Brooks, The role of thermal barrier coating in maximizing gas turbine engine efficiency and lowering CO_2 emissions, Praxair Technology, Inc.: Danbury, CT, (2017).

[18] C. C. Juan, M. H. Tsai, C. W. Tsai, C. M. Lin, W. R. Wang, C. C. Yang, S. K. Chen, S. J. Lin, J. W. Yeh, Enhanced mechanical properties of HfMoTaTiZr and HfMoNbTaTiZr refractory high-entropy alloys, *Intermetallics*, 62 (2015) 76–83.

[19] HC Starck Solutions, C-103 Nb alloy: Properties and applications, (2020). Available at: www.hcstarcksolutions.com/c103-nb-alloy-properties-applications.

[20] C. Huang, Y. Yu, J. Yang, Y. Yan, D. Wang, F. Hu, X. Wang, R. Zhang, G. Feng, Ru/ La_2O_3 catalyst for ammonia decomposition to hydrogen, *Appl. Surf. Sci.*, 476 (2019) 928–936.

[21] M. L. Grilli, D. Valerini, A. E. Slobozeanu, B. O. Postolnyi, S. Balos, A. Rizzo, R. R. Piticescu, Critical raw materials saving by protective coatings under extreme conditions: A review of last trends in alloys and coatings for aerospace engine applications, *Materials*, 14 (2021) 1656.

[22] Ultramet. Ceramic protective coatings. Available at: https://ultramet.com/ceramic-protective-coatings/ (accessed August, 17, 2021).

[23] A. M. Motoc, S. Valsan, A. E. Slobozeanu, M. Corban, D. Valerini ... R. R. Piticescu, Design fabrication, and characterization of new materials based on zirconia doped with mixed rare earth oxides: Review and first experimental results, *Metals*, 10 (2020) 746.

[24] Y. Hu, Z. Yu, G. Fan, Z. Tan, J. Zhou, H. Zhang, Z. Li, D. Zhang, Simultaneous enhancement of strength and ductility with nano dispersoids in nano and ultrafine grain metals: A brief review: *Rev. Adv. Mater. Sci.*, 59 (1) (2020) 352–360.

[25] S. C. Tjong, Z. Y. Ma, Microstructural and mechanical characteristics of in situ metal matrix composites, *Mater. Sci. Eng. R: Reports*, 29 (3–4) (2000) 49–113.

[26] Z. Dong, Z. Ma, L. Yu, Y. Liu, Achieving high strength and ductility in ODS-W alloy by employing oxide@W core-shell nanopowder as precursor, *Nat. Commun.*, 12 (2021) 5052.

[27] J. Reiser, M. Rieth, A. Möslang, B. Dafferner, A. Hoffmann, X. Yi, D. E. J. Armstrong, Tungsten foil laminate for structural divertor applications: Tensile test properties of tungsten foil, *J. Nucl. Mater.*, 434 (1–3) (2013) 357–366.

[28] Q. Wei, L. J. Kecskes, Effect of low-temperature rolling on the tensile behavior of commercially pure tungsten, *Mater. Sci. Eng. A*, 491 (1–2) (2008) 62–69.

[29] N. Durlu, N. K. Çalişkan, S. Bor, Effect of swaging on microstructure and tensile properties of W–Ni–Fe alloys. *Int. J. Refract. Met. Hard Mater.*, 42 (2014) 126–131.

[30] U. R. Kiran, A. Panchal, M. Prem Kumar, M. Sankaranarayana, G. V. S. N. Rao, T. K. Nandy, Refractory metal alloying: A new method for improving mechanical properties of tungsten heavy alloys, *J. Alloys Compd.*, 709 (2017) 609–619.

[31] S. Zhou, Y. J. Liang, Y. Zhu, R. Jian, B. Wang, Y. Xue, L. Wang, F. Wang, High entropy alloy: A promising matrix for high-performance tungsten heavy alloys, *J. Alloys Compd.*, 777 (2019) 1184–1190.

[32] Z. M. Xie, R. Liu, S. Miao, T. Zhang, X. P. Wang, Q. F. Fang, C. S. Liu, G. N. Luo, Effect of high temperature swaging and annealing on the mechanical properties and thermal conductivity of W-Y2O3, *J. Nucl. Mater.*, 464 (2015) 193–199.

[33] Z. M. Xie, R. Liu, T. Zhang, Q. F. Fang, C. S. Liu, X. Liu, G. N. Luo, Achieving high strength/ductility in bulk W-Zr-Y_2O_3 alloy plate with hybrid microstructure, *Mater. Des.*, 107 (2016) 144–152.

[34] H. Dobbelstein, M. Thiele, E. L. Gurevich, E. P. George, A. Ostendorf, Direct metal deposition of refractory high entropy alloy MoNbTaW, *Phys. Procedia*, 83 (2016) 624–633.

[35] Z. E. Tan, J. H. L. Pang, J. Kaminski, H. Pepin, Characterisation of porosity, density, and microstructure of directed energy deposited stainless steel AISI 316L, *Addit. Manuf.*, 25 (2019) 286–296.

[36] S. Rodriguez, A. Kustas, G. Monroe, Metal alloy and RHEA additive manufacturing for nuclear energy and aerospace applications, SAN D2020-7244, Sandia National Laboratories, (2020).

[37] C. Buchanan, L. Gardner, Metal 3D printing in construction: A review of methods, research, applications, opportunities and challenges, *Eng. Struct.*, 180 (2019) 332–348.

[38] J. Murray, Low cost metal additive manufacturing, desktop metal presentation, Roadrunner3D, Albuquerque, NM, February 13, 2020.

[39] B. M. Arkhurst, J. J. Park, C. H. Lee, J. H. Kim, Direct laser deposition of 14Cr oxide dispersion strengthened steel powders using Y_2O_3 and HfO_2 dispersoids, *Korean J. Met. Mater.*, 55 (8) (2017) 550–558.

[40] J. C. Walker, K. M. Berggreen, A. R. Jones, C. J. Sutcliffe, Fabrication of Fe–Cr–Al oxide dispersion strengthened PM2000 alloy using selective laser melting. *Adv. Eng. Mater.*, 11 (2009) 541–546.

[41] T. Boegelein, S. N. Dryepondt, A. Pandey, K. Dawson, G. J. Tatlock, Mechanical response and deformation mechanisms of ferritic oxide dispersion strengthened steel structures produced by selective laser melting, *Acta Mater.*, 87 (2015) 201–215.

[42] M. B. Wilms, R. Streubel, F. Fromel, A. Weisheit, J. Tenkamp, F. Walther, S. Barcikowski, J. H. Schleifenbaum, B. Gokce, Laser additive manufacturing of oxide dispersion strengthened steels using laser-generated nanoparticle-metal composite powders, *Procedia CIRP*, 74 (2018) 196–200.

[43] H. Berns, W. Theisen, *Eisenwerkstoffe Stahl und Gusseisen*, Springer Verlag, Berlin, (2006).

[44] J. Schmelzer, S. K. Rittinghaus, M. B. Wilms, O. Michael, M. Kruger, Strengthening of additively manufactured Me-Si-B (Me = Mo, V) by Y_2O_3 particles, *Int. J. Refract. Met. Hard Mater.*, 101 (2021) 105623.

[45] H. Sun, M. Wang, X. Xi, Z. Nie, Effects of formation of complex oxide dispersions on mechanical properties and microstructure of multi-doped W alloys, *Mater. Sci. Eng. A*, 824 (2021) 141806.

[46] V. Gopan, L. D. Wins, K. A. Surendran, Innovative potential of additive friction stir deposition among current laser based metal additive manufacturing processes: A review, *CIRP J. Manuf. Sci. Technol.*, 32 (2021) 228–248.

[47] S. Rathee, M. Srivastava, S. Maheshwari, T. K. Kundra, A. N. Siddiquee, *Friction Based Additive Manufacturing Technologies Principles for Building in Solid State: Benefits, Limitations, and Applications*, CRC Press, Boca Raton, FL, (2018).

[48] H. Z. Yu, R. S. Mishra, Additive friction stir deposition: A deformation processing route to metal additive manufacturing, *Mater. Res. Lett.*, 9 (2) (2021) 71–83.

[49] D. Gu, D. Dai, W. Chen, H. Chen, Selective laser melting additive manufacturing of hard-to-process tungsten-based alloy parts with novel crystalline growth morphology and enhanced performance, *ASME. J. Manuf. Sci. Eng.*, 138 (8) (2016) 081003.

[50] V. K. Balla, S. Bodhak, S. Bose, A. Bandyopadhyay, Porous tantalum structures for bone implants: Fabrication, mechanical and in vitro biological properties, *Acta Biomater.*, 6 (2010) 3349–3359.

[51] R. Wauthle, J. V. Stok, S. A. Yavari, J. V. Humbeeck, J. P. Kruth, A. A. Zadpoor, H. Weinans, M. Mulier, J. Schrooten, Additively manufactured porous tantalum implants, *Acta Biomater.*, 14 (2015) 217–225.

[52] B. R. Levine, S. Sporer, R. A. Poggie, C. J. D. Valle, J. J. Jacobs, Experimental and clinical performance of porous tantalum in orthopedic surgery, *Biomaterials*, 27 (27) (2006) 4671–4681.

[53] O. R. Mireles, O. Rodriguez, Y. Gao, N. Philips, Additive manufacture of refractory alloy C103 for propulsion applications, AIAA 2020-3500. AIAA Propulsion and Energy 2020 Forum, August (2020).

[54] C. A. Terrazas, J. Mireles, S. M. Gaytan, P. A. Morton, A. Hinojos, P. Frigola, R. B. Wicker, Fabrication and characterization of high-purity niobium using electron beam melting additive manufacturing technology, *Int. J. Adv. Manuf. Technol.*, 84 (2016) 1115–1126.

[55] S. Posen, M. Liepe, Mechanical optimization of superconducting cavities in continuous wave operation, *Phys. Rev. ST Accel. Beams*, (2012), 022002-1–022002-10.

[56] D. Longuevergne, N. Gandolfo, G. Olry, H. Saugnac, S. Blivet, G. Martinet, S. Bousson, An innovative tuning system for superconducting accelerating cavities. Nuclear Instruments and Methods in Physics Research Section A: Accelerators, Spectrometers, Detectors and Associated Equipment, (2014), 7–13.

[57] T. Ries, K. Fong, S. Koscielniak, R. E. Laxdal, G. Stanford, A mechanical turner for the ISAC-II quarter wave superconducting cavities, in: *Proceedings of the 2003 Particle Accelerator Conference, IEEE*, (2003), 1488–1490.

[58] G. Ciovati, G. Cheng, U. Pudasaini, R. A. Rimmer, Multi-metallic conduction cooled superconducting radio-frequency cavity with high thermal stability, *Supercond. Sci. Technol.*, 33 (2020) 07LT01 (7pp).

[59] G. Marinelli, F. Martina, S. Ganguly, S. Williams, H. Lewtas, D. Hancock, Functionally graded structures of refractory metals by Wire Arc Additive Manufacturing, *Sci. Technol. Weld. Join.*, 24 (5) (2019) 495–503.

[60] R. M. Mahamood, E. Akinlabi, *Functionally graded materials*, Springer International Publishing, Cham (2017).

[61] V. Bhavar, P. Kattire, S. Thakare, S. Patil, R. K. P. Singh, A review on functionally gradient materials (FGMs) and their applications, IOP Conference Series: Materials Science and Engineering, IOP Publishing, vol. 229 (2017) 012021.

Index

A

abrasive wear, 144
Accumulative roll bonding, 62, 63, 86
Accumulative roll bonding and folding, 63
activation energy, 3, 66, 85, 92, 97, 108, 119, 135, 149
activator, 25, 119
additive manufacturing, 124, 174–176
adiabatic shear, 45, 126, 134, 135, 165, 166
aerospace, 1, 11, 12, 15, 58, 125
alloying, 12, 15, 16, 18, 19, 25, 44, 53–56, 65, 68, 77, 78, 82–84, 86, 92, 108, 117, 125, 133, 140, 152, 155, 160, 172, 173
amorphous, 15, 56, 58, 78, 79, 81, 82
atomic packing factor, 1
atomization, 57
attritor mills, 54, 68

B

ball-to-powder weight ratio, 54, 84
bimodal grain size, 47, 82, 132, 133
blistering, 25, 124, 144, 149
body-centered cubic, 1
bonding, 1, 2, 9, 15, 18, 35, 36, 39, 62, 63, 86, 95, 101, 107–109, 119, 123, 132, 140, 149, 151, 154, 160
broadening, 78, 79
Burgers vector, 24, 40, 97, 133

C

chemical, 1, 5, 13, 14, 18–21, 65, 67–69, 77, 81, 85, 126, 138, 176
chemical vapor deposition, 5, 18, 65
coating, 6, 11, 13–16, 18, 58, 66–68, 85, 143, 150, 160, 171, 173
coherency, 41, 81, 173
cohesive energy, 1, 21
columnar oxide grains, 160
compaction, 78, 83, 86, 91, 93, 100, 103, 108, 123, 127
compressibility, 91, 123
compressive stress, 5, 39, 105, 156
consolidation, 21, 54, 78, 91, 98, 106–110, 127, 132, 173
contamination, 52, 54, 78, 79, 84, 174, 176
contiguity, 91, 92, 99, 102, 113, 133, 138, 166
conventional pressureless sintering, 83, 91, 92
coordination number, 123
corrosion, 13–16, 18, 19, 39, 126, 166, 167, 171–173
creep deformation, 3, 5, 7, 166
creep rupture life, 7, 8, 12, 25, 47, 177
critical stress intensity factor, 156
crystallites, 19, 20, 85
crystallite size, 44, 51, 78, 84, 96, 107

D

DBTT, 16, 23, 45, 133, 134, 138, 168, 172, 175
de-bonding, 45, 52, 160
defense, 1, 9, 15, 47, 142, 176
densification, 44, 53, 54, 78, 91, 95–98, 100, 103–105, 107, 109, 117–119, 121–127, 131, 133, 134, 144, 151, 152, 168
density, 1–9, 11, 12, 16, 20, 23–25, 35, 38, 39, 41, 45, 51, 53, 54, 58, 59, 67, 68, 78, 81, 83–86, 92, 100, 103–106, 109, 117, 121–127, 131, 132, 143, 149, 150, 155, 165–167, 173, 175, 176
diffusion, 1, 4, 8, 25, 35, 53, 54, 68, 82, 86, 92, 93, 96–101, 108, 109, 117, 118, 122, 125, 149–151, 154–156, 160, 161
diffusion rolling, 86
dihedral angle, 118–119
dislocation, 7, 8, 22–25, 39–42, 44, 54, 58, 59, 64, 77, 81, 85, 98, 103, 104, 109, 122, 131, 133, 135, 138, 141, 168
ductility, 4, 5, 9, 13, 16, 18, 19, 22, 23–25, 44, 45, 47, 51, 53, 77, 82, 101, 103, 126, 132–134, 138, 168, 171, 173, 175
dynamic deformation, 134

E

electric arc melting, 51
electron beam melting, 5, 51, 53, 176
electronic configuration, 1
elongation, 3, 5, 6, 8, 19, 25, 44, 45, 53, 132–134, 141, 144, 166, 173
embrittlement, 16, 19, 44, 93, 124, 138, 140, 167, 171
enthalpy of mixing, 79
equal channel angular pressing, 59
erosion resistance, 166
extrusion ratio, 102

F

face-centered cubic, 1
forgeability, 106

fracture, 3–6, 8, 9, 16, 18, 44, 45, 52, 54, 63, 126, 133, 138, 140–142, 157, 165, 168
free energy, 22, 35, 108, 154, 156
functionally graded materials, 176
furnace heating elements, 166

G

glass-melting furnace components, 166
grain boundary sliding, 5, 24, 44, 95, 124
grain growth, 39, 43, 44, 47, 82, 91, 92, 95, 97, 100, 107–110, 117, 119, 125, 131, 140
green density, 78, 81, 83, 92, 123, 124

H

Hafnium, 1, 8
Hall-Petch relationship, 23
halogen lamp, 166
hardness, 5–9, 13–15, 19, 25, 44, 45, 68, 131, 132, 142–144, 168, 175
heat-resistant alloys, 167
heating rate, 91, 94–96, 99, 100, 110, 123, 124, 127, 131, 165, 168
Herfindahl-Hirschman Index, 171
hexagonal closed packed, 1
high entropy alloy, 45, 51, 67, 68, 79, 81, 85, 104, 135, 136, 141, 143, 149, 171, 174
high pressure sintering, 108, 109
high pressure torsion, 61, 85
HiPIMS, 67, 68
hot isostatic pressing, 100, 131, 143
hydrostatic extrusion, 101, 102
hydrothermal synthesis, 82

I

incoherency, 41
interfacial energy, 42, 43, 119, 160
intergranular, 6, 82, 134, 140
intermetallic, 9, 18, 78, 79, 81, 99, 100, 107, 131, 134, 141, 143, 158
interparticle spacing, 40, 41
intragranular, 82, 104, 144, 168, 173
ionicity, 35, 36
Iridium, 1, 6
irradiation, 16, 18, 19, 44, 45, 68, 84, 126, 138, 140, 151, 153
ITER, 10, 45

J

joule heating, 96

K

kinetic energy penetrator, 9, 11, 16, 126, 133, 134, 165

L

lamp components, 166
laser ablation, 68
laser engineered net shaping, 175
laser metal deposition, 174
lattice misfit, 39, 41–43
lattice parameter, 37, 41, 78, 79
linear law, 155
liquid phase, 45, 91, 92, 98, 100, 102, 117, 119–127, 133, 158

M

magnetron sputter deposition, 67
mechanical alloying, 19, 44, 53–55, 65, 68, 77, 78, 82–84, 86, 125, 133, 172
medical implants, 15, 165, 167, 172, 175, 177
melt extrusion, 56
melting-casting, 20, 51, 53, 68, 77, 173, 176
melting spinning, 56
melting temperature, 1, 4, 11, 13, 21, 22, 35, 54, 58, 77, 122, 167
microcutting, 142
microhardness, 6, 131, 132, 143, 175
microwave sinter forging, 105, 106
microwave sintering, 98–100, 127, 131
milling, 54, 78, 79–84, 92, 102, 125
mixing entropy, 79
Molybdenum, 1, 3
monomodal size distribution, 78
multidirectional forging, 60

N

nano oxide, 43, 44, 77, 125, 140, 160, 167, 168
nanoparticle, 22, 23, 40, 63, 68, 82, 92
nanostructured materials, 19, 20, 23–25, 95, 109
near net shape, 19, 53, 103, 172
Niobium, 1, 4
non-equilibrium, 20, 53–55, 58, 68
nuclear, 1, 7, 9, 12–16, 18, 19, 44, 45, 47, 85, 126, 138, 166, 171

O

Orowan mechanism, 40
Osmium, 1, 7, 15
oxidation, 2–9, 14–19, 25, 39, 47, 51, 53, 58, 91, 106, 125–127, 144, 149–152, 154–158, 160, 161, 165–169, 171, 172, 177
oxidation resistance, 2, 4, 5, 16, 25, 39, 47, 126, 127, 149–152, 154, 156–158, 160, 165, 166, 168, 172
oxidation wear, 144
oxide dispersion, 25, 37, 38, 41, 43–45, 77, 78, 82, 85, 123, 125, 131, 138, 140, 151, 152, 157, 160, 166, 168, 169, 171, 172, 174

Index

oxide scale, 3–5, 8, 39, 149–152, 154–158, 160, 168

P

parabolic law, 155
partial pressure, 6, 7, 16, 154, 155
particle size, 19, 22, 40, 41–44, 54, 57, 60, 65, 68, 77, 78, 81–85, 91, 92, 103, 106, 108, 110, 124, 127, 131, 133, 172
Perovskite oxides, 35
physical vapor deposition, 67
Pilling-Bedworth ratio, 5, 8, 39, 155
pinning pressure, 41, 42, 82
planetary ball mill, 54, 68
plasma-facing material, 9, 85, 126, 166, 167, 176
plasma spraying, 58, 166
plastic deformation, 4, 58, 61, 68, 77, 95, 109, 122–124, 143, 157, 171
plasticity, 9, 53, 102, 138, 152, 158
plowing, 144
porosity, 15, 38, 39, 51, 100, 122, 126, 138, 143, 156, 160, 168, 174, 175
powder forging, 103–105, 127
powder injection molding, 126
powder metallurgy, 5, 53, 104, 123, 126, 134, 141, 172, 173, 176
powder preform, 104
precipitates, 39–42, 104, 134, 138
process control agent, 54, 83

R

rapid solidification, 56
rare earth metals, 39
rare earth oxide, 35, 39, 169, 173
recrystallization, 3, 5, 9, 16, 44, 45, 47, 58, 63, 86, 107, 133–135, 166–168.
refractory metals, 1–3, 7, 9, 15, 25, 37, 39, 44, 47, 51–53, 60, 62, 77, 92, 108, 109, 125, 132, 144, 149, 151, 155, 156, 165, 169, 171–173
residual stress, 9, 103, 171, 174, 176
Rhenium, 1
rocket nozzle, 9, 16, 25, 126, 165, 166, 177
rocket nozzle throat, 165, 166
rolling, 5, 8, 19, 62, 63, 85, 86, 126, 140, 168, 174
rupture strain, 6
Ruthenium, 1, 6

S

scaling law, 23, 92, 93
segregation, 16, 18, 44, 51, 53, 82, 108, 119
selected laser melting, 174
self-sharpening, 45, 126, 134, 165
severe plastic deformation, 58, 61, 68, 77

shear band, 45, 126, 134, 165
shear localization, 166
shear modulus, 7, 8, 19, 24, 40, 41, 59, 97, 133
shock-based consolidation, 106, 108
shrinkage, 44, 91, 92, 97, 105, 110, 118, 119, 122, 126, 127
sinter forging, 105, 106
sintered density, 121, 123–126
sintering, 19, 39, 78, 82, 83, 86, 91–110, 117–127, 131–133, 143, 152, 165, 168, 172
sintering boats, 166
sol-gel, 65, 68, 77, 83, 86
solubility, 8, 12, 16, 18, 19, 20, 51, 53–56, 58, 77, 108, 119, 125, 127, 134, 149, 155, 160, 161, 167
solution reprecipitation, 119
spallation, 25, 144, 152, 156, 158, 160
spark plasma sintering, 93, 94, 124, 127, 131, 132, 165
SPEX shaker mills, 54
sputtering, 3, 52, 67, 167
stacking fault energy, 8, 85
strengthening, 12, 35, 39–41, 44, 45, 53, 83, 104, 132, 174
stress concentration, 44, 132, 134, 144, 171
structural materials, 23, 132
subgrain, 58
sublimation, 5, 11, 16, 18, 25, 39, 66, 121, 149, 152, 154, 155, 161
surface energy, 44, 92, 117–119, 122, 156
sustainability, 3, 11, 58, 64, 127, 155, 160, 165, 171, 173, 176

T

Tantalum, 1, 5
tensile strength, 3, 5, 7, 13, 16, 18, 25, 38, 44, 45, 86, 94, 133, 134, 173
thermal barrier coatings, 173
thermal shock, 9, 11, 45, 52, 95, 160, 166, 168
thermal spraying, 57, 58
thermal stability, 6, 39, 172, 176
toughness, 3, 9, 15, 16, 23, 44, 45, 47, 91, 101, 126, 132–134, 138, 140–142, 144, 157, 168, 171, 176
transgranular, 4, 52, 140, 141
Tungsten, 1, 3, 11, 12, 53, 83, 102, 142, 165
twist extrusion, 64

U

uniaxial pressure, 105

V

vacuum arc melting, 51, 53, 58

vapor pressure, 3–5, 7, 11, 18, 39, 149, 154, 157
volatility, 7, 15, 16, 18

W

wear, 9, 13–15, 25, 44, 79, 126, 142–144, 165, 166, 168, 172, 175, 177
weight loss, 7, 143, 171
wetting angle, 119, 121

workability, 12, 13, 19, 108

Y

yield strength, 8, 24, 25, 44, 102, 109, 133, 142, 175

Z

Zener Pinning effect, 41